Soil and Water Conservation In Semi-arid Areas

Other Related Books

A Manual on Conservation of Soil and Water	Usda
Agriculture and Fertilizer	Bockman Oluf
Agrochemicals in Plant Disease Management	N. G. Ravichandran
Agrotechnology for Dryland Farmaing 2Nd.Ed.	Dhopte A.M.
Agrotechnology for Dryland Farmaing 2Nd.Ed.	A. M. Dhopte
Biofumigation and Solarization for Management of Soil Borne Plant Pathogens	P. Parvatha Reddy
Crop Production in Salt-Affected Soil	I. C. Gupta
Drainage Engineering: Principles and Practices	S. K. Gupta
Fertilizer and Plant Nutrition Guide	Fao
Fertilizer Manual	Ifdc
Fundamentals of Soil Science	A. Rathinasamy
Fundamentals of Soil Science	R. L. Arya
Genesis and Management of Sodic (Alkali) Soils	Gupta S.K.
Glimpse on General Agriculture	P. Laxman. Rao
Indiras Objective Agriculture	R. L. Arya
Managing Manure	Mark Kopecky
Manual on Fundamentals of Agronomy	L. K. Jain
Principles in The Quantitative Analysis of Water, Fertilizers, Plants and Soils	U. S. S. Ramulu
Rainwater Management Theory and Practice	Jat M.L.
Rainwater Management Theory and Practice	M. L. Jat
Salt Affected Soils : Reclamation and Management	R. K. Gupta
Sanrakshit Kheti Ke Antargat Satat Fasal Prabhandhan (Hindi)	K. V. R. Rao
Soil and Plant Analysis	Piper C.S.
Soil and Plant Analysis	C. S. Piper
Soil Chemical Analysis	M. L. Jackson
Soil Conservation Technical Handbook	D. H. Hicks
Soil Guide A Handbook for Uniderstanding and Managing Agricultural Soils	Geoff Moore
The Organic Farming Manual	A. L. Hansen
The Soil-Plant System in Relation To Inorganic Nutrition	M. Fried
Water Quality for Agriculture	R. S. Ayers

Soil and Water Conservation in Semi-arid Areas

Norman W. Hudson

Silsoe Associates
Ampthill, Bedford
United Kingdom

United Book Prints

First Reprint in India, 2018

Published by:
United Book Prints
4806/24, Ansari Road,
Daryaganj
New Delhi - 110 002
Tel.: +91-11-41511055
E-mail: unitedbookprints@gmail.com

Print : 2018

ISBN: 978-93-83692-85-9

Reprint in India

FOREWORD

FAO has been concerned for many years with the development, management and conservation of land resources.

As part of its work in this field the Land and Water Division has, since 1965, produced a series of Soils Bulletins. These have been written with the intention of bringing knowledge to those interested and involved in the development of land resources. To do this, the Bulletins have been aimed at meeting particular requirements as understanding and experience of different aspects of this subject have become available.

In recent years there has been a growing interest in the semi-arid regions of the world. This interest has largely been brought about by the droughts and famines of the 1970s and 1980s that have affected huge areas, and millions of people, in the semi-arid areas of Africa.

What can be done to prevent calamities like these occurring again? There can be no simple answer to this question as a wide range of technical, economic, social and political problems must be examined and resolved.

However, we do know that droughts are very much a part of the natural events in the semi-arid regions - they have occurred often in the past and they will occur frequently again in the future. We also now realize that the terrible effects of these last droughts have been exacerbated by soil erosion and other forms of land degradation which have been allowed to develop over the years in semi-arid regions through mismanagement and poor land use. Degraded land and vegetation is just not capable of withstanding the added demands placed upon them by drought.

But soil erosion can be overcome and the land base restored to a healthy, productive state if sound management and appropriate techniques are applied.

All the answers to conservation in the semi-arid regions could not be given in a short publication such as this - even if they were all known. Exactly what is needed varies from place to place depending on the local climate, soils, vegetation and human requirements.

What has been done in this Bulletin, therefore, is to present methods and techniques which have been tested and proved effective in some part of the world where lack of rainfall is a problem because of amount, distribution or unreliability. Many of these methods and techniques, we believe, have potential for wider use in other parts of the world with similar problems. This Bulletin then is intended as a reference and a guide to those planners and technicians working in the semi-arid areas who are searching for ideas to develop and adapt in their efforts to control land, degradation and introduce sustainable systems of productive agriculture.

The reader is asked to note that owing to the differences in terminology in soil classification systems, the author has used the terms from the orginal works referred to. As different terms mean different things to different people, Annex 2 has been included. This consists of a table taken from Elsevier's Agricultural Compendium (1981), Table 2.7/7, entitled "Summary of FAO/Unesco system of soil classification, with analogues in the US system of taxonomy".

ACKNOWLEDGEMENTS

Assistance is gratefully acknowledged from the many individuals and institutions who have helped by sending ideas, comments, literature and photographs for this Bulletin, particularly:

Staff of the Arab Centre for the Studies of Arid Zones and Dry Lands (ACSAD), Damascus, Syria; P.E.V. Charman, and staff of the Soil Conservation Service of New South Wales, Australia; Dr.. M.T. Dennet, Dept. of Agricultural Botany, University of Reading; Dr. Robert M. Dixon, Imprinting Foundation, Tucson, Arizona; Dr. H.E. Dregne, International Centre for Arid and Semi-arid Land Studies, Texas Tech. University, Lubbock, Texas; Dr. S.A. El-Swaify, College of Tropical Agriculture and Human Resources, University of Hawaii; Prof. Michael Evenari, Hebrew University of Jerusalem; Dr. Dwayne H. Fink, US Water Conservation Lab., Phoenix, Arizona; Barney Foran, Division of Wild Life and Rangelands Research, CSIRO, Australia; Dr. S.A. Gavande, Soils and Reclamation Consultant, Austin, Texas; Dr G. T. Gowon, Federal Dept. of Agricultural Land Resources, Kaduna, Nigeria; G.M. Hallam, DRSPR, Mali; Dr B. Heusch, Saint Mury-La Tour, France; W. Hoogmoed, Agricultural University of Wageningen, The Netherlands; Dr. Hans Hurni, Soil Conservation Research Project, Addis Ababa, Ethiopia; Staff of the Intermediate Technology Development Group, Water Group; Intermediate Technology Publications; Staff of the International Centre for Agricultural Research in the Dry Areas (ICARDA), Aleppo, Syria; Staff of the International Crops Research Institute for the Semi-arid Tropics (ICRISAT), Hyderabad, India; Dr. Emlyn Jones, Chiredzi Research Station, Zimbabwe; C.D. Kagoro, The Institute of Agricultural Engineering, Zimbabwe; I.A.F. Laing, Dept. of Agriculture, Western Australia; Dr. H. Lal, EMBRAPA, Petrolina, Brazil; J. G. Lewis, WAPAC, World Bank; Prof C. Mathieu, Faculty of Agronomic Science, University of Burundi; Prof. W.G. Matlock, College of Agriculture, University of Arizona; Dr. A. Millington, Dept. of Geography, University of Reading; I.R. Melville, Conservation Commission of the Northern Territory, Australia; Dr. V.V. Dhruva Narayana, and staff of, the Central Soil and Water Conservation Research and Training Institute, Dehradun, India; Dr. W.E. Ormerod, Padworth, Reading; Staff of Overseas Development Institute, London; Staff of Oxfam; Dr. A.von Peter, International Potash Institute, Switzerland; Prof. J. De Ploey and Dr J. Poesen, Catholic University of Leuven, Belgium; Staff of "Institut de Recherches Agronomiques Tropicales et des Cultures Vivrieres" (IRAT), Montpellier, France; R.G. Richards, Conservation Commission of the Northern Territory, Australia; Dr. E. Roose, ORSTOM, Montpellier, France; Prof. C.W. Rose,

School of Australian Environmental Studies, Griffith University, Queensland, Australia; Dr. D. Rosenzweig, Director of Soil Erosion Research Station, Natanya, Israel; T.F. Shaxson, FAO, Brazil; D. Sims, FAO, Portugal; C.W. Steley and W.F. Buddee, Overseas Projects Division, South Australian Dept. of Agriculture; Dr. M.A. Stocking, School of Development Studies, University of East Anglia; Dr. R.K. Sivanappan, Tamil Nadu Agricultural University, Coimbatore, India; Dr. C. Valentin, ORSTOM, Abidjan, Cote d'Ivoire; H. Vogel, German Agency for Technical Cooperation, (GTZ); Dr. T.J. Willcocks, Overseas Division, NIAE, Silsoe, Bedford.

Many technical staff in FAO have also been most helpful, particularly D.W. Sanders, Officer-in-Charge, Soil Resources, Management and Conservation Service, who proposed the bulletin and helped define its scope and objectives, and the officers in that service for technical advice. The FAO Remote Sensing Centre who provided the satellite imagery.

The research for such a wide-ranging subject was greatly helped by support from the library of Silsoe College - data-base searches by the Librarian Bryan Morgan, retrieval through Inter Library Loan by Mrs Betty Laing, and all the staff helped find material in the College Library.

On the production side, many photographs have been kindly loaned, and the source of each is acknowledged in the captions. Photos without a listed source are from the author's collection. Most of the diagrams were drawn by Helen Trayner. I am most grateful to Linde Ovington Lee for ably taking the inumerable drafts and changes through her word-processor, and to Hazel Tonkin, AGLD Editor and Meetings Officer, for editorial help and advice, and for putting it all together.

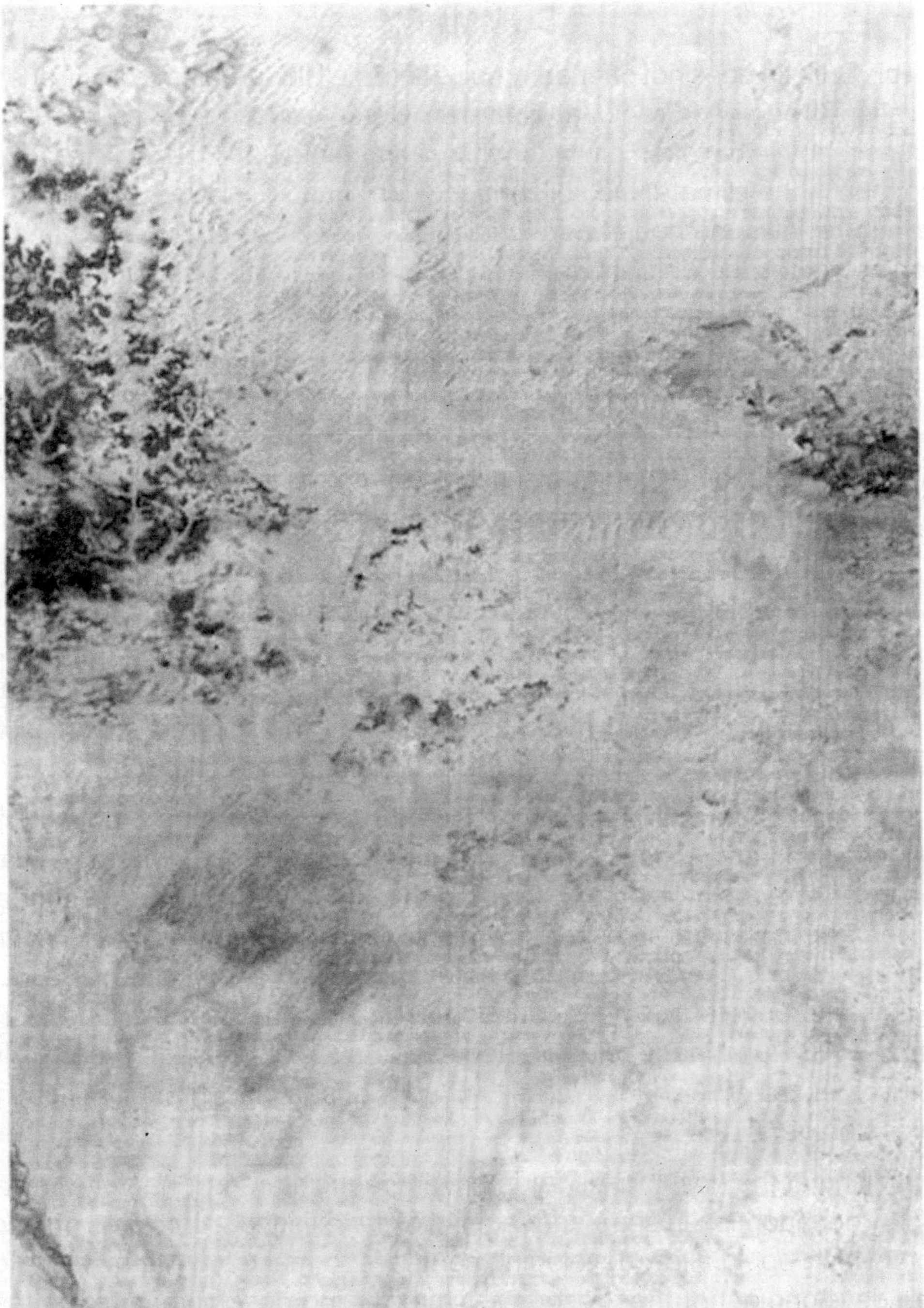

Frontispiece: Satellite imaginery showing overgrazing in the Sahel. The hexagonal shape is the Ekrafane ranch in Niger, where grazing is controlled. The surrounding white area is denuded of vegetation by overgrazing (NASA)

SUMMARY

This Bulletin does not offer easy solutions to all the problems of soil and water conservation in semi-arid regions. There is no storehouse of tested methods and techniques ready to be taken off the shelf for immediate application. The conditions vary too much - the climate, the soil, and the social and economic factors. Instead, this Bulletin reviews methods and techniques which have been tested and found useful somewhere, and which might be suitable for use in other conditions.

We have not attempted to define semi-arid areas, and some of the examples come from arid or sub-humid regions. The objective is to make the bulletin relevant anywhere that rainfall is a problem because of amount, distribution, or unreliability. We have omitted the questions of salinity and alkalinity, wind erosion, and mechanization. Neither is there a discussion of the political, social, and economic issues, although in many cases these are as important as the technical problems.

The first three chapters are introductory, and outline the scale and importance of the problem, the difficulties and the possibilities for improvement. The bulletin argues strongly, and presents evidence, that drought is part of the natural order in semi-arid areas, and that the recent disasters of degradation and famine in Africa result from misuse and mis-management of the natural resources which reduced the region's ability to cope with the additional stress of drought.

Chapter 2 starts with a review of the extent of the erosion problem and the pressures on semi-arid ecosystems resulting from increasing human and livestock populations. There are problems which could be overcome, such as the shortage of information from agricultural research, which has in the past been mainly directed towards areas of higher agricultural potential. This has resulted in a lack of technology appropriate to the needs of sub- sistence farmers and nomadic pastoralists. The unchangeable problems are also discussed, particularly the variability of rainfall. The wide diver- sity of soils brings problems of soil chemistry and water storage.

Looking at the possibilities for improving agriculture in Chapter 3, a case is made for wider adoption in semi-arid regions of the existing techniques of resource inventory which would allow developing countries to make long-term plans for the optimum development of their resources, maximizing development of better land and minimizing stress on the marginal areas. To help the adoption of improved farming methods, Farming Systems Research is required to provide a better understanding of the motivation of subsistence farmers, semi-nomadic cultivators and pastoralists. The slow rate of adoption of promising new ideas is associated with failure to understand the farmers' thinking.

The first of the four technical chapters deals with soil conserva- tion under the headings Principles, Biological Soil Conservation, and Mechancial Conservation Works. It is argued that in semi-arid regions, with inevitable low production, much of the conventional approach is inappropriate because it has been developed in very different conditions. Expensive works using machinery are seldom justified, nor are the conser- vation tillage methods which are the backbone of biological soil conservation in developed countries with a more favourable climate. The requirements in semi-arid areas are that biological conservation through improved farming should show short-term benefit to the farmer, and that this need not necessarily be in terms of increased yield, for he or she may be more concerned with increasing the reliability of yield, or decreasing the labour requirement. Similarly, the emphasis should be on mechanical works which are simple, low-cost, and can be laid out and constructed and maintained with only small inputs of technical advice.

Chapter 5, called Water Harvesting, deals with methods to increase the amount of moisture stored in the soil profile or where there is some small movement as surplus run-off. The choice of methods of water conservation is difficult because the desired objective may change from one season to another. In a dry area it may be sensible to increase surface storage to improve crop yield in most years, but in a wet year this could cause waterlogging and reduce the yield. On the other hand, a drainage system may have the objective of increasing run-off from areas usually too wet, but also have the undesired effect of exaggerating the effect of a drought. The principle of many methods is to make use of marginally inadequate rainfall by concentrating run-off from an uncultivated part of the land onto a cultivated part which then receives enough moisture to grow a crop. The general term for methods of this type is "rainfall multiplier". Examples are conservation bench terraces, developed and widely used in the United States of America, and a number of simpler variations known as contour furrows, microcatchments, or contour bunds. The practice of ridging and tied ridging is discussed in some detail because there is a large amount of experience of this method in many countries, with varying degrees of success. The broad bed and furrow method is being increasingly adopted on the heavy black clays of India, closely associated with improved animal-drawn equipment which brings the method within the reach of peasant farmers. Other methods of localized surface storage are the semi-circular catchments and trapezoidal catchpits becoming popular in Burkina Faso and Kenya, which are variations on traditional methods of surface pitting.

Under the section Water Spreading there is a discussion of various forms of localized use of surface run-off, either occurring naturally from hill areas or after it has been diverted and collected in channels. Several ancient examples of this type of run-off farming are described from the American Indians, Tunisia, and Israel. There are also more recent examples from Kenya, Pakistan, and the Yemen Arab Republic. Inundation methods are where floodwaters are impounded and retained long enough to saturate the soil so that a crop can be grown on the stored moisture. These range from small-scale examples from the Sudan to Brazil to the much larger schemes in semi-arid areas of India, some of which have been used for hundreds of years. The diversion and spreading of floods and spate flows without storage are discussed with examples from Pakistan, Yemen Arab Republic, and several countries in Africa. Sub-surface drainage is felt to have limited application in semi-arid regions, but several examples of simple surface drainage systems are described.

Methods where surface run-off is collected and stored in dams, tanks, or cisterns for later use are described in Chapter 6 on Water Harvesting and Use. The amount of run-off is critical to such schemes, so there is a discussion of ways of treating the catchment to improve run-off. This includes simple methods such as removing stones and vegetation, shaping the catchments as with the roaded catchments and flat-batter dams of Western Australia, and chemical treatments such as deflocculants and waxes, and going on to sealing the surface with waterproof membranes. Practical guidance is given on the design and construction of small earth dams and weirs. Sand dams are discussed, where water is stored in the pore spaces of sand retained by a weir, a method which may be particularly useful in hot climates because surface evaporation losses are reduced. Off-stream storage is described, including the traditional 'hafirs' of Arabic-speaking northern Africa, the tanks of India, and the small reservoirs used in Western Australia for stock watering. There is a discussion of methods for reducing seepage losses through the floor of storage basins, and evaporation losses from the surface, although the available solutions to both these problems are frequently too expensive for general use. This chapter ends with a discussion on groundwater, covering methods of recharge and extraction from wells, boreholes, and horizontal wells, including the ancient qanats of north Africa and western Asia, and modern drilled horizontal wells in the western USA.

Chapter 7 discusses applications of water conservation to grazing land and to trees and shrubs. The problem of overstocking and overgrazing was briefly mentioned earlier among the difficulties and problems in semi-arid areas, and is developed here in more detail. The evidence of satellite imagery proves conclusively that the degradation of the Sahel and of Namibia was man-made and caused by overgrazing, and not the result of drought. Examples are shown from Niger in the Frontispiece and from Namibia in this chapter. There is no doubt that national herds of livestock are increasing and numerical examples are quoted from Swaziland, and from the Mambilla Plateau on the border of Nigeria with Cameroon. The decreas- ing mobility of nomadic or semi-nomadic pastoralists is discussed, quoting the example of Botswana where the two main causes are settlement, and the provision of permanent water supplies at the cattle posts by replacing the previous natural water supply with boreholes. Two examples of successful group livestock management schemes are mentioned, from Niger and from eastern Senegal. Examples of good management are taken from the western USA and compared to the traditional management of the grazing by linking it to water supplies in the Butana region of the Sudan. Techniques for improving the grazing by reducing surface run-off are discussed with examples of large-scale mechanized schemes of creating furrows or pits. Small scale methods more suitable for Africa are semi-circular hoops, low banks of stone set out on the contour, and small basins formed by digging with hand hoes.

In semi-arid areas the landscape is always a mixture of woodland, bush, grazing land and arable land, and trees or shrubs are always a part of the ecosystem. They supply many needs; fuel, browse, construction materials, food crops, and cash crops, and usually a combination of more than one of these. Depletion of the natural stock of trees and shrubs is one of the indicators of excessive pressure on the land leading to an increasing rate of degradation. Ethiopia is the model for a study of the progressive phases of degradation starting with depletion of the tree growth, and ending in a massive exodus. In many countries the increasing scarcity of fuelwood is becoming as critical as the scarcity of food, and in this context, the need to expand the production of trees and shrubs is clearly a major challenge in the semi-arid areas.

The usual techniques for afforesation are seldom appropriate in semi-arid areas because of the unreliability of the rainfall. The methods suggested are variations of the rainfall multiplier approach discussed earlier for cropland. They range from a small basin in which a tree seedling is planted, perhaps with small channels to lead extra water to the basin, up to the 250 m^2 microcatchments developed in the Negev desert in Israel. Other variations of this principle are quoted from several countries in Africa. Agroforestry is considered outside the scope of this Bulletin, but a list of references is provided.

TABLE OF CONTENTS

LIST OF PLATES

LIST OF FIGURES

LIST OF TABLES

1. INTRODUCTION

1.1 THE SCOPE OF THIS BULLETIN

This bulletin does not offer easy solutions to all the problems of soil and water conservation in semi-arid regions. There is no storehouse of tested methods and techniques which can be taken off the shelf for immediate application. The conditions vary too much - the soil, the climate, the social factors such as land tenure, availability of mechanization and of labour, the place of livestock - the list is endless. This bulletin is therefore a review of techniques which have been tested and found useful somewhere, and which might be suitable for use in other conditions. Some of the methods are at the stage of being promising ideas which need more testing.

Since the bulletin covers such a wide range of geography, people, and climate, it will not meet the requirements of every reader. There will be too much detail in one place, not enough in others. In particular there might be more emphasis on the Mediterranean region, where the winter rain- fall regime gives a situation very different from the semi-arid regions with summer rainfall. We are also conscious that we have not done justice to the work in the Francophone countries of North and West Africa.

We have not attempted to define semi-arid areas, nor have we been concerned whether areas should be classified as arid or semi-arid. The working definition of areas which the bulletin is intended to help is anywhere that rainfall is a problem because of amount, distribution, or unreliability. We have not discussed all the problems of semi-arid areas, and particularly have omitted the questions of salinity and alkalinity, and wind erosion, and mechanization. A subject as wide as this could be arranged in chapters in a number of different ways, so we have tried to set up sign posts where a method or technique is discussed in more than one chapter. Figure 1.1 is a generalized map of arid and semi-arid areas.

The bulletin does not go into detailed design of all the practices because the circumstances of soil and climate make each application different, so we have tried to be more concerned with principles. Neither have we discussed the practicability of the methods reported in terms of manpower requirements, nor in terms of costs and benefits. Whether it makes sense to apply a method will depend mainly on the urgency of the problem; if a method will keep people alive the benefit/cost ratio is not the most important criterion.

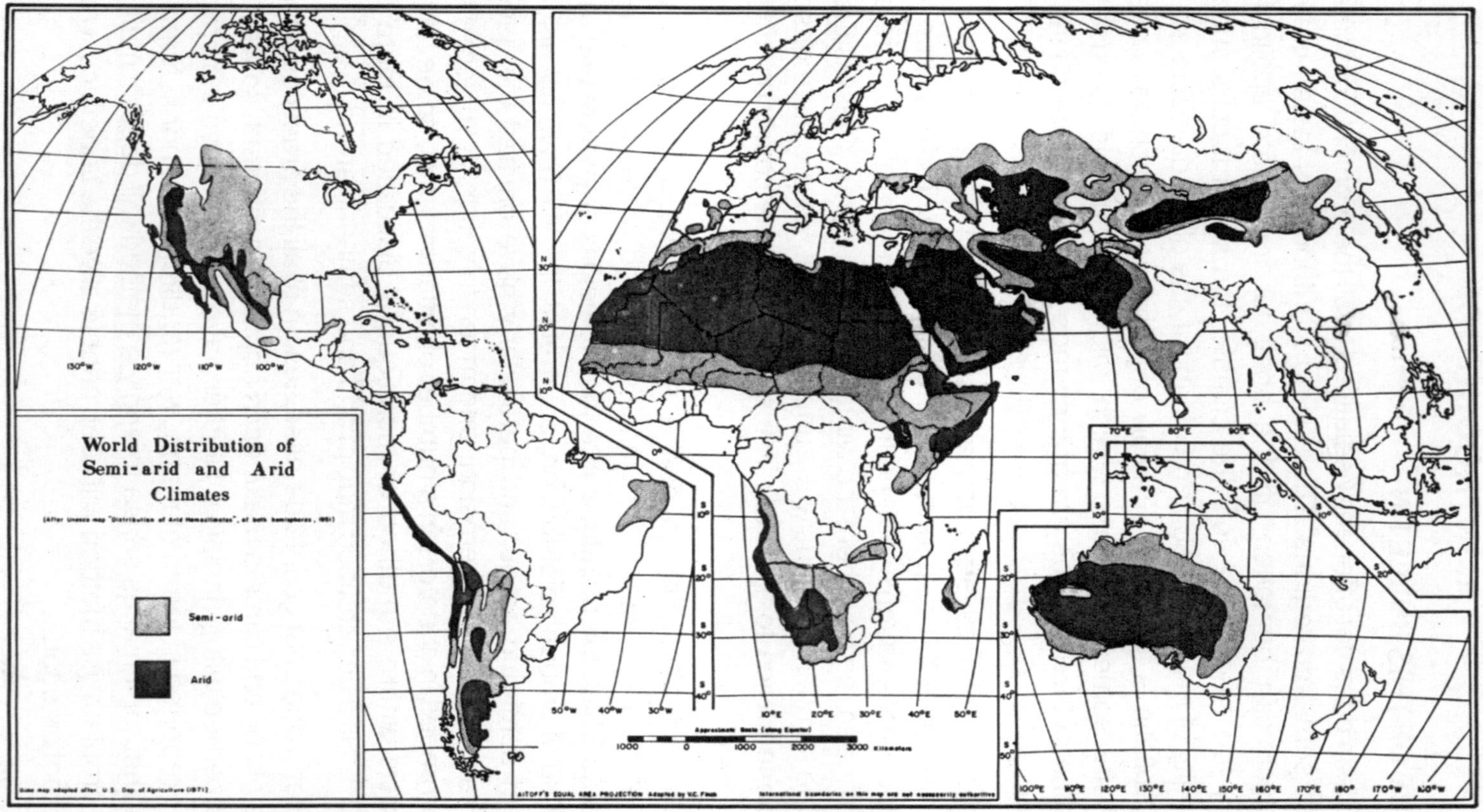

Figure 1.1 Generalised map of arid and semi-arid regions

The book does not discuss the political issues, such as land tenure and social systems, nor does it discuss political structure of government and the social structure of land use. This is not to imply that these aspects are not important. The consensus of many recent analyses of soil and water conservation programmes is that past programmes have been biased towards top-down projects with insufficient attention to the needs, aspira- tions and abilities of the peasant farmer who, as a result, has not been sufficiently involved. It is perhaps significant that many of the ideas and techniques discussed in this bulletin are either indigenous or have been developed together with the farmers by one-man projects operated by non-government organizations.

The other feature which comes through strongly when studying successful projects, is the importance of helping farmers by training them in the use of simple equipment. Examples are teaching the use of the water-tube level to lay out works on the contour in Burkina Faso, or the use of the line level for similar purposes in Kenya. Another important non-technical contribution can be to help the organization of local com- munities to manage their affairs more effectively. In India there is a centuries old tradition of local groups to manage water affairs. There are the Pani Panchayat for general water management, and the irrigation com- mittees, Pukka Warabundi (formally structured), and Katcha Warabundi (informal). In other countries, political and social change is leading to new organizations of Cooperatives or Peasant Associations, and guiding, assisting, or training these groups may be essential for the implementation of any soil and water conservation programmes.

1.2 THE IMPORTANCE OF THE SEMI-ARID REGIONS

Agricultural development naturally takes place first on the best land. Whether at the scale of the individual farm or a whole country, the tendency is to use the best land first. When there is a need to increase agricultural production it is usually directed to maximizing production in the areas which have the best potential. But as demand increases for the products of the land - food, fuel, shelter, and clothing - it is necessary to make increasing use of land which is less suitable for agriculture, or land in less favourable climates. People concerned with agricultural planning, or development, or production, must all pay more attention to the semi-arid regions.

Attention has been sharpened by the widespread droughts in Africa in the early nineteen-seventies, and again in the mid-eighties, as shown by the surge of conferences and workshops in the last ten years, and the explosion of aid programmes in semi-arid Africa.

It is futile to expect magic solutions to the problems of semi-arid regions. We cannot alter the unreliability of the rainfall, nor the fact that unpredictable rainfall and the occurrence of droughts are inevitable. Neither can centuries of abuse and mismanagement of the land and the people be corrected overnight. Short-term results to reduce starvation through improved yields is only part of the story; a programme to win the confidence of subsistence farmers should be planned to a timespan of at least five years, and plans to direct the attitudes of governments more towards land use will need much longer.

What this bulletin tries to do is to put ideas and techniques into a large array of labelled pigeonholes, from which technicians can select components to build into a project or a programme.

1.3 DROUGHT AND CHANGE

The place of droughts in relation to food supplies was admirably expressed by Sir Joseph Hutchinson in the preface to the Royal Society Symposium of 1977, when he said:

> "These problems (widespread droughts of the early nineteen seventies) were not created by drought. Drought is but the dry extreme of the natural climate variability of semi-arid areas. The problems arise from the pressure of human and domestic animal populations on limited resources, and they have been aggravated by the rapid explosion of populations in recent years....
>
> "It is in the nature of these marginal areas that agriculturally disastrous seasons are within the normal range of climatic variations. Such seasons must be accepted as something to be foreseen and planned for, and not as an "act of God" to be met by international charities. The difficulty of planning for drought conditions and hence of preventing famine has been increased by the great successes of science and technology in the last quarter century in the control of parasitic and nutritional disease.
>
> Improvement in health has led to great increases in human and animal populations, and these have been maintained by fuller exploitation of the environment. Thus the pressure at the margin has increased, and what was a shortage that could be met by encroachment on unused resources becomes a famine because no unused resources remain."

The all too common attitude towards drought is typified by the Australian outback farmer who is reported to have explained his problems to the extension officer because "we have not had a normal rainy season for 25 years".

Hutchinson's comment about increasing pressure at the margins is reinforced by Ormerod (1978) who is concerned about the problems which may inadvertently arise from economic development in dry areas, and argues that "One of the most important factors operating to increase aridity in West Africa is the economic demand which has stimulated the growth of herds and of arable farming which compete for land in the arid range areas."

Ormerod suggests that the conventional view that demand is best satisfied by increasing local production may increase the rate of agricul- tural degradation and goes on to say:

> "There is a particular danger in stimulating economic advance in desert areas and particularly in linking nomads to the monetary system and I make the plea that, before barriers to development ... in this remote and fragile part of the world are broken down, more information should be obtained about the dynamics of nomadic grazing and of the possible effects that the extension of their operations might have upon the environment and the climate ...."

He also urges the development of quantitative techniques which could be used to assess the capacity of soils to withstand exploitation, to provide a measure of the limits to which economic development can be pressed without causing ecological degradation.

A third commentary is the recent and thoroughly researched review by Sinclair and Fryxell (1985) who present a powerful argument for the view that the ecological disaster of the Sahel is primarily man-made.

Proof of the damage caused by overgrazing comes from satellite imagery, and examples are shown from Niger in the Frontispiece, and from Namibia in Plate 7.2

1.4 POLITICAL AND SOCIAL CHANGE

A recent change has been the increasing political interest in many countries to promote more agricultural use of areas of low or erratic rainfall. In some cases the main reason has been political change associated with the move from colonial status to independence, and an example is Zimbabwe. In colonial times, agricultural development and agricultural research were primarily directed towards maximizing the production on the highly mechanized large farms on the better soils in the regions with most reliable rainfall. Since independence there has been a reversal of this policy and the emphasis is now placed on development where there is the greatest need rather than the greatest opportunity, i.e. the small-scale subsistence farmer on poor soils in areas of low and unreliable rainfall. Another powerful factor which has led to increasing emphasis on production from the dry areas is the simple necessity to increase total national production. Examples of this are the massive development of grain production in the south-east of USSR, and the increasing attention to production in the low rainfall areas of Kenya.

Yet another reason for opening up areas previously considered too dry for arable farming has been the development of new techniques such as mechanized minimum tillage farming which in New South Wales is leading to the spread of arable farming to the west.

However increasing government interest in dry regions is not universal, and in some countries production is decreasing, because there is apparently no political will to develop the dry regions. A correspondent reports that this is the situation in Morocco, Algeria, and Tunisia. In broad terms, each of these countries was self-sufficient in food grains with average yields of around 1000 kg/ha. Morocco has maintained this average yield, but now has to import 20 percent of its food requirements. In Tunisia the yield has fallen to 800 kg/ha and 40 percent of the food requirements are imported. In Algeria the yield has fallen to 600 kg/ha and food imports are running at 60 percent (FAO 1985).

In some countries a reduction in the population has made it impossible to maintain the conservation works necessary to sustain agricul- ture. This is the case in the Yemen Arab Republic, and also applies to some of the Mediterranean countries on the north African coast.

2. THE PROBLEM OF AGRICULTURE IN THE SEMI-ARID REGIONS

2.1 SOME GENERAL PROBLEMS

2.1.1 The Extent of Erosion

Most semi-arid regions suffer severe rainfall erosion, often more than in the humid tropics, because the rain has a high erosive capacity which is more damaging because, of the reduced vegetative protection Past erosion in Mediterranean lands is well documented by Lowdermilk (1975) and a recent survey shows that the position continues to deteriorate. Some examples are Morocco, with 40 percent of the country's total land area exposed to erosion; Turkey with more than 50 percent damaged; Greece with 21 percent suffering moderate erosion and 46 percent severe erosion damage.

The non-Mediterranean semi-arid regions also report rainfall erosion problems; in North America, the USA and Mexico; in South America, the north-east corner of Brazil, and much of the south-east of the continent; in Africa, the Sahel and the south-west (e.g. Botswana); then there is a huge belt stretching east from Africa through Turkey, Iraq, Iran, and Afghanistan, with one arm extending into eastern USSR and China, and the other arm south of the Himalaya through Pakistan to India. In Australia, most of the states have severe rainfall erosion problems in areas of low rainfall.

Wind erosion can be serious in the temperate semi-arid regions and examples are the central steppes of Asia (eastern USSR, Mongolia, and China); in North America the north-western States of the USA, and the central prairies of Canada. In the semi-arid tropics the picture varies. Some countries surrounding the arid Sahara report desertification with increasing wind erosion. On the other hand, the opinion has been expressed that in many cases the semi-arid regions receive more soil blown by the wind from deserts than they lose from rainfall erosion. This view is supported by the existence of vast areas of wind-deposited loess soil in semi-arid regions.

2.1.2 Differences Between Semi-arid Regions

The only thing common to all the semi-arid regions is low rainfall. Everything else varies so much that it is not practical to think in terms of searching for universal solutions. They may have temperate or tropical climates, rainfall in summer or winter seasons, and hence totally different forms of agriculture.

Even when the climate is similar there are other significant differences as shown when Charreau (1977) considered the possibilities of transferring the technology developed at ICRISAT in India to countries in West Africa. The main features of the climate are the same in both regions: a short rainy season, intensive rainfall interspersed with unpredictable droughts, and highly variable rainfall during the wet season. However, in India the duration of the rainy season is longer, and the curve of potential evapotranspiration is unimodal, while it is bimodal in West Africa. Non-typical periods without rain are most likely to occur in the middle of the rainy season in India, but at the beginning and the end of the season in West Africa. Both regions have, as their main soils, red alfisols and black vertisols, but there are differences which have important implications for cropping, cultivation, drainage, and fertility status. In addition there are major sociological differences, particularly the density of population which in India is very high at 200 inhabitants per km^2 compared to 10-60 per km^2 in semi-arid West Africa. This leads to a very different approach to agricultural production. In West Africa the objective is to maximize production per worker, whereas in India it is more important to maximize the return per unit of land. There are also differences in the style of land use. In Asia families have farmed the same land for generations; in Africa many people are nomadic or seasonally migrant.

2.1.3 Shortage of Information

Previously the emphasis of agricultural research and development has been on making the best use of good soils and good climates and little attention was given to marginal environments. One result of this uneven spread of effort is a lack of information about the less favoured semi-arid areas. This shows up in every technical discipline. In the semi-arid regions there are fewer soil surveys to show what is available, and less research on the physical and chemical properties of soils to show their capablities and problems. There are fewer agrometeorological stations, and less analyses of the results, although because of the variability there really should be more of both.

There has been less research on crops. Most countries with a combination of humid and arid conditions have concentrated on the humid areas, sensibly because this is likely to give the quickest and most cost-effective return. Major international research for dry areas is recent, with the establishment of the International Crops Research Institute for the Semi-Arid Tropics (ICRISAT) in 1972 and the International Centre For Agricultural Research in the Dry Areas (ICARDA) in 1977. Selecting or breeding cultivars for particular semi-arid climates shows promise, but has a long way to go (Hall et al. 1979).

There is also a lack of information on the social and economic side. The farming systems, and the economics of agricultural production, have been studied less, partly as a result of the lower interest, partly because of the physical difficulties of large areas and fewer roads. Also the farming systems are complex because of:

· combinations of settled agriculture and nomadic or semi-nomadic livestock, or seasonal transhumance;

· cultural complexities of communal ownership or tribal grazing rights;

· cultural and ethnic boundaries may be different from the political boundaries.

2.1.4 Lack of Technology

A combination of lack of interest, low research commitment, and the complexities of the problem results in a shortage of technology which can be applied to improving agriculture in semi-arid regions. Jones points out that, in Africa, studies of water management techniques such as tillage and tied-ridging have several times shown great promise, but have been too brief, or too local, or restricted to experiment stations, and as a result they "generate neither the depth and breadth of understanding of the technique (and the limitations imposed on it by soil, implements, econo- mics, etc.) nor the impetus and general interest that might sustain it through a long programme of wider testing, adjustment, and retesting under practical farming conditions" (Jones 1984). Research in semi-arid regions needs a long time span to reduce the effect of seasonal variations, and coverage over a wide area to reduce the effect of local variations.

Another difficulty is putting technology into widespread operation after it has been developed and tested. In India, agricultural production is unstable in over three-quarters of the agricultural areas because of inadequate or erratic rainfall. The Drought-prone Areas Programme has made considerable progress over twenty years, but there is still a gap between the research results and practical application to farming (Johl 1981). Improved technology for the vertisols, developed at ICRISAT, has been tested and validated, and is being increasingly taken up. But to get it applied on the whole 5 million hectares of vertisols within 20 years would require a growth rate of 43 percent per year in the rate of adoption (ICRISAT 1986).

An important reason for the poor adoption of new techniques is the inability of the subsistence farmer to take risks. The essence of farming is trying to improve the odds in the gamble against weather, pest, and disease. The peasant has no risk capital to gamble with, so his whole strategy is geared to safety. He would rather use a low-yielding variety which gives some yield every year than an improved variety which will give an increased yield most years, but none at all in the bad year. Even if the chance of an increased yield is nine years out of ten, it is still not an acceptable gamble for the small subsistence farmer. In the tenth year, the year of failure, his family will starve. It is neither stupidity nor lethargy when he sticks to his old variety, it is accepting the realities of life. Figure 2.1 illustrates this argument by comparing the yield of a local variety of a cereal and a hybrid high yielding variety. The point that the subsistence farmer cannot afford to take risks applies equally to many other possible changes which may be promising but uncertain.

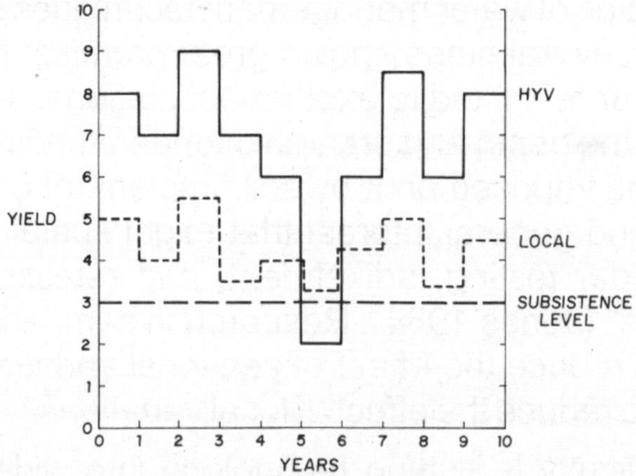

Figure 2.1 The high yeldin variety (HYV) has a higher average yield than the local variety, but is unacceptable if it occasionally yields less than the minimum required for subsistence

2.2 PRESSURES

The human population of semi-arid regions is increasing like any other region as a result of natural increase, but there are other factors. There is a general migration from more densely settled areas, in some cases encouraged by national policies of land allocation or for improving produc- tion in the drier regions. It has also been suggested that the increase of settled agriculture in areas previously occupied by nomadic pastoralists has been accompanied by a higher growth rate among the settled population than the nomads. Increases in the livestock population are sufficiently well documented in many countries for us to be able to recognize a trend of steady increase, usually an upward exponential curve. Some examples are given in Section 7.1.1. One factor has been the success of veterinary science (referred to in Chapter 1), another is that the demand for meat products has risen dramatically in some countries, and a case in point is Nigeria. In some cases the development of water supplies has been another factor leading to more livestock, and more pressure on grazing, lands.

An interesting example is the sheep population in Iraq as quoted by Thalen (1977), who suggests that a remarkable increase in the sheep population resulted from a combination of the introduction of mechanical equipment for drilling wells and the adoption of four-wheel-drive vehicles which greatly increased the accessibility of remote regions. The national herd was estimated at 4 to 5 million in the nineteen-thirties and rose to 10 or 11 million in the nineteen-sixties, leading to such overgrazing and degradation that the population fell back to 6 million in the early nineteen-seventies. Similar examples of over usage following improved water supplies are reported from the Sudan (Adams 1975) and from other countries in the Sahel (Franke and Chasin 1981).

Increased pressure on the natural resources also leads to loss of flexibility. There are no reserves which can be called upon in the bad year. This applies at the scale of the individual farmer whose meagre reserves of food and grazing are soon used; it applies equally at the level of the village or tribe who in the past have traditionally held emergency grazing in reserve; it applies at the national level as grain stocks are depleted, and at the regional level when it is no longer pos- sible to use the safety valve of major movement of livestock. Examples of traditional safety valves are Botswana, where a poor season is the occas- ion for the large movement of stock out to the east of the country, and the Sahel where there is a general movement southward, in both cases with a reverse migration as conditions improve. In western Sudan the effect of recent droughts has not been so severe because the people were able to move south with their livestock to wetter areas which are only sparsely populated.

Political events may provide restraints on traditional movements of this nature by restricting the movement across national frontiers. On the other hand, political troubles can also lead to problems caused by the movement of refugees.

The general trend of change in land use has been quantified in the case of Mali. Between 1952 and 1975 savanna grazing went down from 89 to 63 percent of the farmed land while cropping rose from 6.6 to 11.1 percent. At the same time degraded land increased from 4 to 26 percent (ILCA 1980).

To summarize, there is a wide range of evidence to show that the natural resources of the semi-arid regions are suffering increasing damage as they face increasing demands and increasing pressures.

2.3 CLIMATE

2.3.1 Rainfall

The two key features which adversely affect agriculture are the low amounts of rainfall and the unreliability. Adding to the problem, the two are linked, in that the variability becomes greater as the mean rainfall decreases. Data from western Asia and North Africa show this in Figure 2.2. The low quantities of rainfall lead to several contraints. Cropping may not be possible at all, or possible only with the use of special techniques such as fallows, or furrows to collect more rain. The range of suitable species will be limited, and in the semi-arid tropics with summer rainfall it is usual to grow maize where the rainfall permits, then sorghum where the rain is not sufficient for maize, and millet where the rain is not sufficient for sorghum. Farming strategy must therefore be directed towards minimizing any loss or wastage of rainfall.

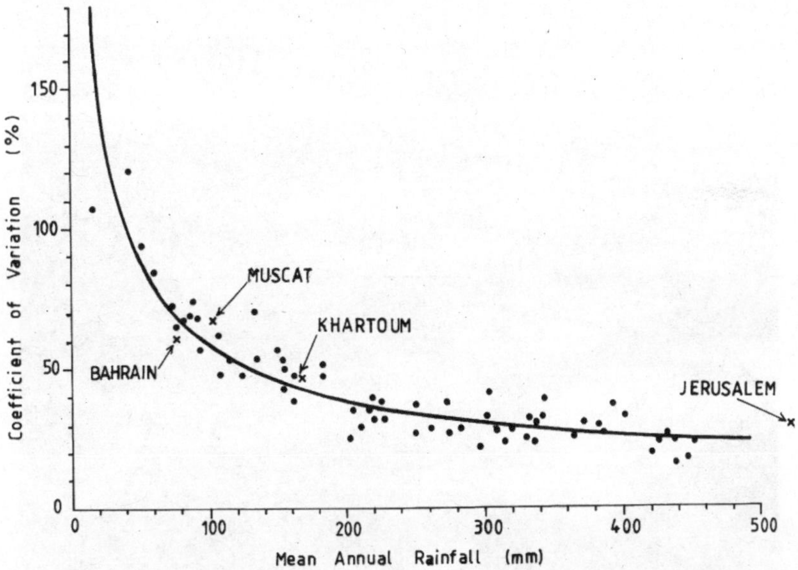

Figure 2.2 Coefficient of variation of annual rainfall as a function of mean annual amount (from FAO Irrigation and Drainage Paper 37, Arid Zone Hydrology, 1981)

Areas of low annual rainfall may still receive heavy individual storms causing flood and erosion. Table 2.1 from Rainey (1977) shows the large differences between mean annual rainfall and figures for the exceptional wettest month. Examples of devastating floods and erosion resulting from abnormal storms in Queensland, Australia, are shown in Plates 2.1 and 2.2.

There is a general association between mean annual rainfall and the annual erosive power, but this results mainly from the fact that a higher annual rainfall includes a greater number of storms, not that the indivi- dual storms are more severe. The high-intensity, short-duration convective rainfall dominant in the tropical semi-arid regions has fairly stable intensity/duration relationships, and these are independent of the long- term rainfall at a particular location. For example, an important parameter for the prediction of erosion is the maximum fall in a 30-minute period and this has been shown to be independent of location and of mean annual rainfall both in Mediterranean lands and the semi-arid tropics (FAO 1981, p 13). On the other hand, some areas experience low-intensity frontal-type rains, usually in the winter season. Where most precipitation occurs in winter, as in Jordan and in the Negev, low-intensity rainfall may represent the greater part of annual rainfall, and rainfall erosion is generally less in these circumstances, although sensational floods can still occur.

Plate 2.1 Erosion damage from an abnormal storm, Queensland, Australia (Queensland Dept. of Primary Industries)

Plate 2.2 Erosion damage from an abnormal storm, Queensland, Australia (Queensland Dept. of Primary Industries)

TABLE 2.1

HEAVY RAINS IN ARID LANDS (mm)

		Mean annual rainfall	Wettest month on record	Next wettest month
Sudan	Wadi Halfa 1937-60	3	29(July)	14(October)
	Abu Hamed 1931-60	17	110(July)	48(August)
Algeria	Aoulef 1931-60	11	25(January)	20 (Aug & Nov)
Chad	Faya-Largeau 1941-60	16	48(June)	31 (August)
South Yemen	Aden 1941-60	39	55(November)	47(September)
Pakistan	Jacobabad	99	330 (July)	129 (August)

Source: WMO (1967) Climatological normals for 1931-60.

The increasing variability with increasing dryness was shown in Figure 2.2, and another way of illustrating it is to look at the standard deviation of annual rainfall. In temperate climates, in nineteen years out of twenty, annual rainfall is between 75 and 125 percent of the mean. In the semi-arid tropics, at mean annual rainfalls of 200-300 mm, the rainfall in nineteen years out of twenty typically ranges from 40 to 200 percent of the mean, and for annual rainfalls of 100 mm the range widens and is 30 to 350 percent of the mean (FAO 1981, p 13).

There is usually a considerable positive skewness of the distribu- tion of annual point rainfalls. This means that there are more annual values below the mean than above it, and the modal value is less than the median value, and both the mode and the median are less than the mean (FAO 1981). There is also the question of short-term spatial variation which is common when the rainfall is mainly of the convective small-cell type. For example, Sharon (1972) showed that in Israel the rain may be in small cells of up to 5 km diameter, so that in a watershed of a few hundred km^2 only 20 percent of the area may receive rain. Over the longer term, variation of the rainfall regime, within an area of generally similar rainfall, can be sufficient to influence crop management, as shown by a study of three sites in Syria by Keatinge et al. (1985), and at six sites in Botswana by Dennett and Rodgers (1986).

In other words, the general unreliability of rainfall from year to year is compounded by the variations from place to place and from storm to storm. Forecasting or prediction is therefore also difficult. Modern statistical techniques allow the calculation of the probability of events important to agriculture such as the occurrence of planting rains, or of drought spells at particular times during the season. But these are essentially strategic tools, helpful to governments for planning policies, or offering guidance on the probability of being able to carry out crop management practices at particular times. There are promising possibil- ities of providing seasonal forecasts for a region, based on links between large-scale features of the atmospheric circulation and rainfall, for example a study of Fiji by Dennett et al. (1978), but such forecasts can only be general and will show considerable local variability. What the farmer really needs is advice to help him in short-term tactical decisions, but this is more difficult. Studies by Dennett et al. (1983) found no dependence between rainfall at different times within the rainy season; that is, events early in the season do not provide any guide on what is likely to happen later in the season. On the other hand, in Kenya, Stewart and Hash (1982) were able to assign a rainy season into one of three categories of predicted total rainfall, according to the time and early amount of the early rains, and to do this sufficiently early to modify management decisions. But a similar analysis of Botswana rainfall data revealed no linkage between early rainfall and seasonal total (Sinclair 1985).

There is thus some opportunity for tactical mid-season changes. In another example, Rainey (1977) observes that nomadic graziers respond quickly and effectively to seasonal variations in rainfall by moving herds to where their bush telegraph intelligence system tells them that rain has fallen. Mid-season changes of cropping patterns are more difficult. Theoretically it should be possible to plant a preferred long duration variety if the rains start on time, but switch to a short duration crop if the rains are late. There are some cases of this technique being used, for example in India and in Swaziland, but there are serious logistic problems of maintaining duplicate seed stocks and of distributing them at the required time.

2.3.2 Temperature and Wind

Extremes of temperature can cause problems at both ends of the scale. The wheat growing areas of Kazakhstan, USSR, are only free from frost for 100-120 days, and cropping is only made possible by day and night ploughing, planting as early as possible, and harvesting in autumn before the crop is fully mature Botswana in Southern Africa experiences surface temperatures so high that germination can be inhibited, and it is possible that one reason for reduced yields after ridging is an increased soil temperature in the ridges.

Wind erosion is likely to be a threat whenever soil dryness and wind occur together. This combination is frequent in the semi-arid tropics, and can also occur in temperate climates when seasonal dryness coincides with seasonal winds. An example from the Columbia Basin Irrigation Project in Washington State, USA, in early spring, is shown in Plate 2.3.

Hot dry winds have three main effects. They reduce the effective- ness of rainfall by evaporation from the soil surface; they increase evapotranspiration from the leaf area of crops, increasing the risk of moisture stress; and the surface of stored water suffers from high evaporation loss.

Plate 2.3 Wind erosion on an irrigation project in Washington State, USA, in early spring (USDA - SCS)

2.4 SOIL

The diversity of soils in semi-arid regions is immense, and two examples will serve to indicate the range of soils and soil-related problems. In a review of the soils of the semi-arid tropics, Kampen and Burford (1980) show that the semi-arid tropics are mainly in the developing countries of Africa (70 percent of the total) and south-east Asia (mostly in India). Latin America and Australia each contain about 10 percent. Eiqht soil orders are represented, of which alfisols and aridisols are more than half of the total. Alfisols are the largest order and cover about 32 percent of the African SAT and about 38 percent of the Asian SAT. Vertisols cover 6 percent of the total area but 25 percent of semi-arid India.

Another example is Jones (1984), explaining the diversity of soils in the low rainfall areas of southern Africa, "the soils vary widely from very extensive areas of arenosols (sands and loamy sands) in Botswana, Zimbabwe and Mozambique, to smaller but potentially important areas of vertisols (heavy clays) scattered widely over the zone; and from highly leached and acid acrisols, ferralsols, and nitosols in Tanzania (mainly) to neutral, alkaline, and salt-affected soils in Botswana (fluvisols, solonchaks, and xerosols and solonetz)."

Physical properties which adversely affect agriculture on these soils are equally variable. Some examples are:

· low rates of infiltration leading to high run-off and hence less effective utilization of the rainfall;

· surface crusting is widespread in semi-arid regions and may be the primary reason for low infiltration (Valentin 1985). Apart from the reduction in infiltration, surface crusting may hinder the emergence of seedlings to the point where it has to be included as a factor in capability classification, as in Zimbabwe;

· dense packing of soils may reduce infiltration and also increase the draught requirement for tillage operations, as in the 'hardveld' sandy loams in Botswana;

· the deep cracking of vertisols can lead to increased loss of moisture by evaporation, and to cultivation problems;

· many semi-arid soils have low moisture-holding capacity, especially shallow or sandy soils. The shallow alfisols in India are an example.

Water storage in the root zone may be limited by a) low intrinsic moisture-holding capacity of sandy soils and b) the limitation on root pro- liferation at depth by natural soil hardness. Management techniques to prevent run-off may cause ponding and over-wet conditions particularly at early stages of growth. Due to erratic rainfall and low storage, even com- plete infiltration early in the season may not avoid moisture stress later.

Chemical soil problems include low fertility which may be inherent or caused by leaching or by past soil erosion. Breman and Vithol (1984) report that, in the Sahelian rangelands, the low fertility of soils, especially in nitrogen and phosphate, is often more of a limiting factor than low and irregular rainfall. Salinity and alkalinity are another major problem but outside the scope of this book.

There are a number of specialized forms of erosion which are asso- ciated with excessively high erodibility and which are more frequent in semi-arid than in humid regions. Tunnel erosion or piping is one of these and often associated with badlands geomorphology (Bryan and Yair 1982). This may be associated with dispersing soils or soils with a calcium/sodium imbalance. Geological erosion can be severe as a result of a combination of high erodibility, heavy storms, and poor vegetative cover, as in the example of Plate 2.4.

Plate 2.4 Severe Geological erosion in a semi-arid climate (Fairchild Surveys)

2.5 THE FARMING BACKGROUND

All the problems of developing countries are present in the semiarid regions, but usually intensified by the low levels of production, which lead to little or no investment capacity on the farm, or low development of the infrastructure by government. The general picture is rather depressing with declining production and increasing degradation, but examples show that techniques exist to overcome the problem.

i. In the semi-arid Pacific north-west of the USA, successful wheat production has been achieved on mean annual rainfall of 240 - 400 mm but this is mainly winter rainfall, and the system requires substantial inputs of machinery, capital, and fertilizer (Bolton 1977).

ii. Turkey is another success story (Kronstad 1981), where the annual yield of wheat rose from 8 million tonnes in 1965 to 16 million tonnes in 1976 as the result of a programme which included:

· research programmes on tillage, weed control, and breeding;

· a substantial increase in the national extension and training programme;

· advanced training overseas for research and extension workers. In this case also there were substantial inputs but they consisted of training, research and extension, so the example is more relevant and applicable to developing countries.

iii. In Australia, reasonable levels of yield are being maintained through the use of clover or medic rotations. The only fertilizer used is superphosphate at 60 to 100 kg/ha. Grazing of the legumes and stover has to be carefully controlled. This system has been successfully demonstrated in Libya, and has potential for other areas.

iv. In summer rainfall areas there are promising developments of dry-land farming techniques coming from ICRISAT, although farmer acceptance is still limited.

In many countries a serious constraint is the shortage of energy available for cultivation. In the third world, the poorest farmers can only dream of the day when they may be able to own a pair of oxen, and millions more have some animals but not enough to work their land effectively. In summer rainfall areas there is always the problem that in spring, when there is the greatest need for draught power, and ploughing has the highest draught requirement, the animals are in poor condition after coming through a hard dry winter. It is common in Africa to find that after the rains start, the animals have to be fed up on the new grass for a month to get them into condition, although early planting is one of the most basic requirements for good crop production. In other countries the critical shortage may be manpower. Migration of adult male labour to more rewarding occupations in towns or in neighbouring countries has a serious effect on agriculture in many countries. In Africa, the gold mines and industry of South Africa have drawn away so much labour from neighbouring Lesotho and Swaziland that crop production is materially reduced. Another recent market for unskilled labour has sprung up in the oil-rich Middle East. Migration from densely populated countries like India and Pakistan does not affect their agriculture, but in other cases there are problems. In the Yemen Arab Republic a well-developed system of terracing is collapsing because of inadequate labour to maintain it (Vogel 1985). A similar situation exists in several countries of North Africa where migration to the labour markets of Europe is adversely affecting agriculture.

There are also all the usual problems of inadequate inputs. Fertilizer and improved seed are either not available or too expensive for the subsistence farmer. He has inadequate investment capital, and coopera- tives and sources of credit are usually undeveloped, and particularly difficult for semi-nomadic peoples.

There are also the social problems arising from the difficulty of effectively managing communal grazing land; and the reluctance to make investments of cash or labour in land which is held on a temporary or insecure lease. The infrastructure is usually less well developed in semi-arid regions because there has been both less demand for roads, water supplies, and marketing outlets, and also because the low level of produc- tivity does not justify development of these facilities. There are usually problems of distribution of seed, chemicals, and fertilizer, and these have recently been thrown into prominence by the inability to distribute food provided by aid agencies in countries suffering famine. While most governments have policies directed towards the improvement and development of agriculture, these are not always pursued as vigorously as they might be. It has been suggested that in several countries in Africa the government policy of reducing the national herd to relieve the pressure on degrading grazing land is not happening because so many of the stock are in very large herds owned by powerful individuals. Another difficulty for governments is how to reconcile the political wish to hold down the cost of staple foods and yet at the same time pay a fair price to the farmers. At the level of international relations, it is alarming that there are so many political conflicts in semi-arid regions where the struggle against nature is bad enough without being made worse by further conflict.

3.ASSESSING THE POSSIBILITIES FOR IMPROVING AGRICULTURE

To plan improved agriculture, or even to attempt to halt or slow down the effect of soil degradation, we must know the facts which determine what is possible and what is practical, what the physical resources allow, and which of these options can be made to work in human terms.

3.1 PHYSICAL POTENTIAL

It is remarkable that so many countries, whose future development is going to depend mainly on agricultural production, have not taken the important step of compiling an inventory of their available natural resources, and a classification into agro-ecological zones of different potentials. Such surveys can be done using methods according to the scale. The FAO programme of regional agro-ecological studies uses massive computa- tional capacity to combine the 1:5 000 000 World Soil Map with climatic data to give maps of agricultural potentials. The same method can be applied to countries, with mapping at 1:1 000 000. Studies at this scale are useful for very broad strategic planning but need to be followed up by more detailed studies (FAO 1978).

Another style of resource inventory is the series of natural resources studies by the Land Resources Development Centre (LRDC) which combine air-photo interpretation and ground survey with existing country data on geology, soils, and climate, and may be at a scale suitable for the whole of Northern Nigeria (LRDC 1968), or at 1:250 000 for the tiny country of Lesotho (LRDC 1968). An effective system has been developed in Zimbabwe which could serve as a model for most developing countries. The agro-eco- logical survey of Zimbabwe was completed and mapped at 1:1 000 000 in the nineteen-fifties, and has been progressively refined and improved to the present "natural regions and farming areas" which is an essential tool for the government strategic planning of agricultural policy for Zimbabwe. Australia provides good examples of detailed surveys at catchment level (Aldrick et al. 1978). The French service for overseas research (ORSTOM) has developed in Tunisia an ecological classification with 22 "ecological systems", suitable for Mediterranean semi-arid zones in North Africa (Floret 1981).

Erosion also needs to be surveyed and recorded, at different scales according to the degree of detail required. Landsat and the constantly improving satellite imagery, can be used for broad assessments (Pickup and Nelson 1984), or air-photo interpretation for more detail (Keech 1985), in both cases with field work to establish ground truth. As well as studying present erosion we should collect information on soil conservation. What has been done in the past, even if it has not been successful, can provide useful information on mistakes to be avoided. Also we need to study conservation methods in case they can also be used in other situations.

The fact that semi-arid regions have not been intensively developed has led to a generally low level of information on soils and climate. Rainfall measurements tend to be few and far between whereas the strong spacial variability means that the available data may be not representa- tive. Similarly, soil surveys and studies of the physical and chemical characteristics of soils are in general much weaker than in areas of more reliable rainfall.

Because of the wide variation of conditions in semi-arid regions, the results of research and development in other countries need to be interpreted with caution, but in recent years there has been a greater interest in farming drier areas and so any review of existing information should include information from research programmes, particularly current and recent programmes where the results may not have been widely dissemi- nated. In the most common situation of existing knowledge and information being inadequate, programmes of research and development must include and bring together both component research within disciplines, and research on farming systems both on-station and on-farm.

3.2 SOCIAL POSSIBILITIES

There is little doubt that in the past much competent component research has not been successfully transferred and adopted by the farmers because it did not fit into the farming system. It is increasingly recog- nized that a complete understanding is required of all the social and economic ramifications of the farming system in order to guide component research into topics and techniques which will be applicable. Both col- lecting the information and understanding it is particularly difficult in the case of farming systems in semi-arid regions. The usual difficulty in collecting data from farmers, such as scepticism of official intentions, and reluctance to disclose financial information, are compounded by the complexities of the system. For example, in commercial farming for profit, it is fairly straightforward to find out the process of taking management decisions and who the decisions are taken by. In tribal or village commu- nities, taking decisons can be a very lengthy and complicated operation. The facts which are being sought are complicated and may be unfamiliar to the people conducting the survey. Rights of tenure and rights of access, e.g. to water supplies, to grazing, to fuel, or cultivation rights or gathering rights are usually complicated. In nomadic or semi-nomadic live- stock management even ownership may not be straightforward; for example, individual ownership of animals within a communal herd, or ownership of the animal but not its offspring, or conditional ownership with reversion if a marriage breaks up, and many other complications unfamiliar to western agriculture. There can be no doubt that studying the whole of the agricul- tural system is fundamental to any consideration of the possibilties for agricultural development. Techniques for obtaining the required information have been greatly improved in recent years, by Mann (1974), Chambers (1981), and Collinson (1981).

Many of the strategies for improved cropping and methods to make more effective use of the rainfall, discussed in Chapters 5 and 6, are most likely to be applicable to individual land holdings or to small groups. This probably accounts for the popularity and success of small development schemes operated by Non Government Organizations (NGOs), for example the Oxfam project in Burkina Faso, or the project by War on Want in Kenya, which are discussed in Section 5.2.4. The classic phrase of Schumaker "small is beautiful" certainly seems to apply to most development projects in semi-arid areas. There are of course some examples of large-scale run-off farming operations, such as the ancient systems in the Negev or the inundation tanks or khadins of northern India. One such tank near Bharatpur has an impounding bund 19 km long, inundates 4100 hectares, and irrigates another 4800. These schemes are described in Section 5.3.3.

The slow rate of adoption of what appears to be a wholly satisfac- tory new method was discussed against the background of extension in India in Section 2.1.4, but there are also examples of speedy adoption, such as the recent rapid spread in Botswana of the use of medium horsepower tractors and borehole syndicates. A study of why new ideas sometimes move slowly and in other cases take off rapidly would be an interesting field of study. One problem of the adoption of methods which attempt to change the rainfall/run-off/crop usage pattern is that the rainfall varies too much to form the base of a reliable strategy. The water collection device which is required in a dry year, is counterproductive in a wet year when drains are required, and vice versa.

In the long term there are opportunities for improving the physio- logy of plants for use in semi-arid areas. The selection and breeding of short-maturing varieties is well established and has resulted in some significant successes, and there are other aspects not so well developed such as increasing the transpiration efficiency. The question of more effective use of available water is discussed further in Section 4.2.4.

4. SOIL CONSERVATION

4.1 PRINCIPLES

4.1.1 The Extent of Erosion

The lower rainfall in semi-arid areas compared with that in humid climates does not mean a corresponding low level of soil erosion by water. Indeed rainfall erosion can be higher in semi-arid areas than in any other climatic zone. This is partly because the rainfall of semi-arid areas has a high proportion of convective thunderstorm rain of high intensity and high erosive power. It is also because there is poor protective vegetative cover, especially at the beginning of the rainy season.

Some of the soils common in semi-arid areas are particularly vulnerable, either because they have poor resistance to erosion (high erodibility), or because of their chemical and physical properties. An example from Mexico is illustrated in Plate 4.1 For example, alfisols suffer a particularly high loss of productivity per unit loss of soil (Stocking and Peake 1985). Gully erosion can be severe in semi-arid climates and the benefit/cost of gully control needs to be considered. Successful but expensive gully conservation like the Australian example shown in Plate 4.2 might not be suitable for third world countries.

4.1.2 Soil Conservation and Water Conservation

There are always strong links between measures for soil conservation and measures for water conservation, and this applies equally in semi-arid areas. Many measures are directed primarily to one or the other, but most contain an element of both. Reduction of surface run-off by structures or by changes in land management will also help to reduce erosion. Similarly, reducing erosion will usually involve preventing splash erosion, or formation of crusts, or breakdown of structure, all of which will increase infiltration, and so help the water conservation.

4.1.3 Integrated Programmes

The approach by soil conservationists in the 1980s is moving away from using mechanical works and structures in soil conservation programmes paid for by a government or a donor-funded project. An example is the increasing awareness of the ineffectiveness of terracing programmes alone. Also, we are moving towards the view that the only effective programmes are those which have the full support of the people. The subsistence farmer cannot afford to respond to philosophical or emotional appeals to care for the soil, and this means that conservation measures must have visible short-term benefits to the farmer. For the subsistence farmer the benefit he would most appreciate might be increased yields per unit of land, or perhaps better production per unit of labour, or perhaps improved reliabi- lity of yield.

The idea of working together in groups on tasks which require a big labour force is well-established in many countries, particularly for planting or harvesting. The practice can be successfully extended to conservation works. The advantages are:

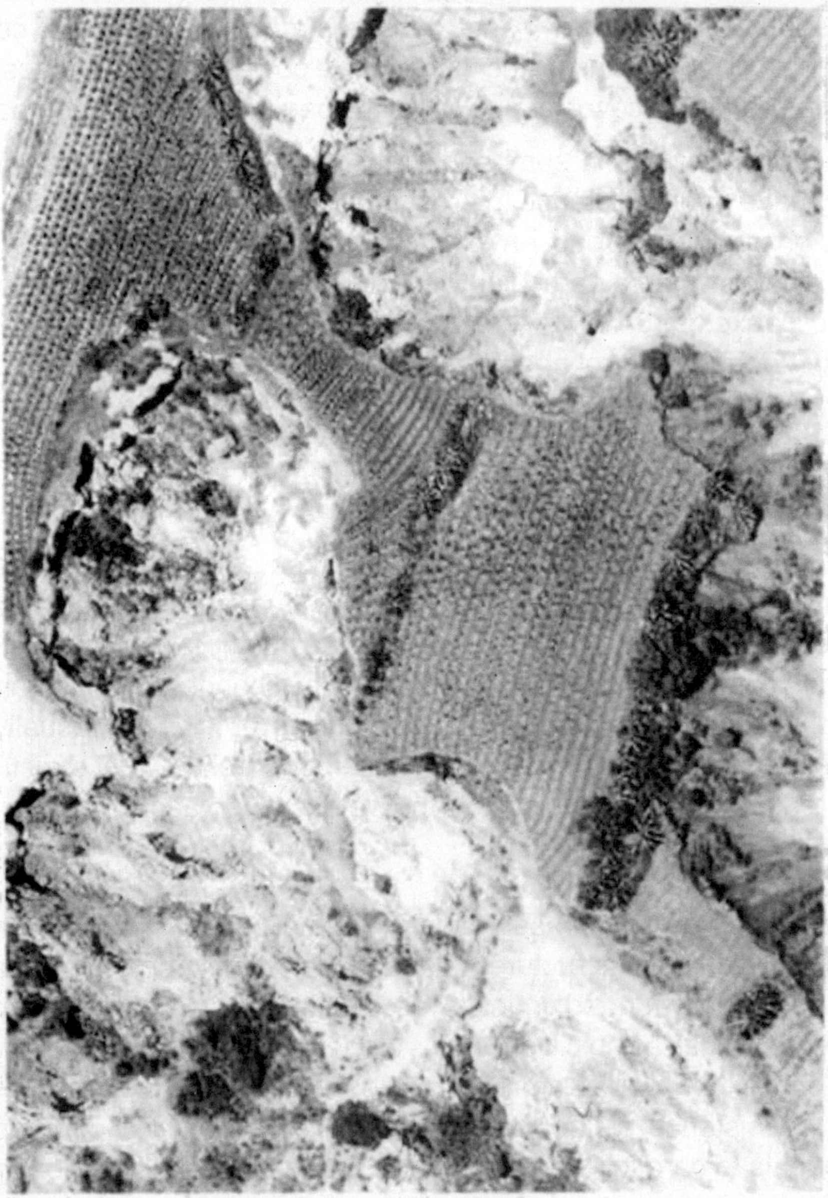

Plate 4.1 An erodible soil devastated by water erosion in Mexico (Roberto C. Coutino)

BEFORE

AFTER
One season later
after battering
and chisel seeding

Plate 4.2 An example of expensive gully control from Australia (R.J. Noonan, Soil Conservation Authority, Victoria)

- a village or group can tackle jobs too big for an individual or family;

- it generates a sense of community care for the land;

- work groups are a good forum for extension workers to encourage improved farming methods (Plate 4.3).

Plate 4.3

A self - help group in Kenya (Curt Matzon)

4.1.4 Design Requirements

If we accept the argument that soil conservation must be cost- effective to be acceptable to the farmer, then the low value of production from semi-arid soils means that only cheap and simple solutions are appro- priate. On a fertile soil with good rainfall it may be sensible to invest a lot of labour or money in sophisticated schemes for controlling the run- off, but not in semi-arid areas with low and unreliable yields. It follows that attempts to eliminate soil erosion completely may be unrealistic, and that some level of erosion may have to be accepted, and also some risk' of soil conservation measures failing. An example of a realistic approach to the risk of failure are the flood diversion dams built in the People's Democratic Republic of Yemen for spate irrigation schemes. Each end of the diversion is built of stone, or nowadays concrete, with a simple earth centre section. It is accepted that the earth section will be destroyed by big floods but it is cheap to repair or replace (Thomas 1982). To upgrade the design and construction so that they could withstand the 25-year flood would increase the construction effort beyond what the farmers can provide. This same approach should be applied to all mechanical conservation programmes in semi-arid areas.

4.1.5 Relevant Technology

Many conservation programmes have failed because the technology was inappropraite, or misapplied, or because they did not take account of the social situation and did not involve the people. The record of soil conservation in north Africa is striking. Heusch (1985) concludes that the large conservation programmes in Algeria, Morocco and Tunisia, from 1950 to 1975, were based on inappropriate technology imported from the totally different conditions of the United States, and the whole effort was a mistake which should not be repeated. Similar criticisms have been levelled at the GERES project in Burkina Faso.

4.2 BIOLOGICAL SOIL CONSERVATION

4.2.1 Conservation Tillage

This umbrella term can include reduced tillage, minimum tillage, no-till, direct drill, mulch tillage, stubble-mulch farming, trash farming, strip tillage, plough-plant (for details see Mannering and Fenster 1983). In countries with advanced soil conservation programmes, particularly the USA and Australia, the concept of conservation tillage is the main theme of the recommendations for cropland, and it is also being taken up quickly in other areas, for example southern Brazil. The application is mainly in mechanized high production farming with good rainfall, or for the control of wind erosion where there is large-scale mechanized cereal production. It is less applicable to low input level crop production, or subsistence agriculture.

The principles are equally effective in any conditions - to maximize cover by returning crop residues and not inverting the top soil, and by using a high crop density of vigorous crops. Conservation tillage also has the advantage of reducing the need for terraces or other permanent struc- tures. However there are several disadvantages which hinder the application of conservation tillage in semi-arid conditions:

- dense plant covers may be incompatible with the well-tested strategy of using low plant populations to suit low moisture availability;
- crop residues may be of value as feed for livestock;
- planting through surface mulches is not easy for ox-drawn planters although there may be no problem with hand jab planters.

Surface manipulation such as ridging is discussed in Chapter 5.

4.2.2 Deep Tillage

One of the reasons for low yields in semi-arid areas is the limited amount of moisture available to crop roots. The available moisture will be increased if the rooting depth is increased and it has been shown that in some cases deep tillage can help, for example on the dense sandy soils (luvisols) in Botswana (Willcocks 1984). Reviewing many studies of experi- ments of depth of tillage on alfisols, El-Swaify finds varied results; deep tillage is beneficial for some crops but not all, and on some soils but not all. Also deep tillage requires greater draught power which is usually in short supply in semi-arid areas.

Ripping or subsoiling can be beneficial, either to increase the porosity of the soil, or to break a pan which is reducing permeability. The deep placement of fertilizer can also be used to encourage more rooting at depth, but again the application of this technique to subsistence farming will be difficult.

4.2.3 Conservation Farming

Like conservation tillage, this title covers many different farming techniques. It includes any farming practice which improves yield, or reliability, or decreases the inputs of labour or fertilizer, or anything else leading towards improved land husbandry, which we have defined as the foundation of good soil conservation.

Sometimes there is a long history of traditional farming and soil conservation practices which have been tested and developed over periods of time which are long enough to include all the likely variations of climate. These traditional practices should give the best long-term result, bearing in mind that the farmer's interpretation of 'best' may be based on reliability rather than maximum yield. But the semi-arid areas are changing rapidly, and the traditional patterns may be no longer relevant. As Jones (1985) says "while tradition may incorporate the wisdom of centuries of practical experience, it may also be inappropriate where recent demographic pressures have already compelled changes - for instance, the abandonment of bush fallowing or migration onto different types of soil or into more arid areas. There is also the point that the agricultural scientist very often still lacks the recipe for certain success; and you cannot require farmers to adopt new practices that are only 50 percent successful." Possible new techniques should have the same basic characteristics as traditional practices, they should be easy to understand, simple to apply, have low inputs of labour or cash, and must show a high success rate i.e. a high rate of return.

Some of the techniques are:

Farming on a rade is well established in India (Swaminathan 1982).Cultivations and planting are done on a gentle gradient, sometimes together with graded channel terraces. This encourages infiltration but permits surplus run-off at low velocities. Sometimes this may be combined with simple practices to encourage infiltration such as returning crop residues. This seldom provides a complete solution because of the problem of disposal of the surface run-off when it does occur.

Strip cropping is most useful on gentle slopes, where it may reduce erosion to acceptable levels without any banks or drains.

Rotations are another well established and simple practice. The object may be to improve fertility by the use of legumes or to help control pest or disease. In the semi-arid parts of Australia a successful practice is to alternate a cereal crop with a free seeding self-regenerating annual forage legume such as subterranean clover or medicago. Trials of adapting this system in Tunisia are reported by Doolette (1977).

Fallowing is well established and successful in some circumstances but not others. In the drier wheat lands of Australia, a bare fallow in summer is used to build up soil moisture before sowing the winter wheat which receives only barely adequate rainfall. The practice is particularly useful on cracking clay soils. There is a risk of erosion taking place during the summer when high-intensity summer thunderstorms fall on the bare soil (Walker 1982). In East Africa, using this method on sloping land has a high risk of erosion (Pereira et al. 1958), but on gentler slopes in Botswana good results were reported by Whiteman (1975). The practice is not universally successful, partly because subsistence farmers may fail to keep the fallow completely free of weeds, and it is unlikely to appeal in uni-modal rainfall if the result will be twice as much grain half as often. In Syria, ICARDA studied the effect of fallows on moisture conservation in a barley/fallow rotation at six sites with annual rainfalls varying from 260 to 350 mm. At less than 260 mm there was no increase in stored moisture, and farmer-managed fallows had little effect up to 300 mm, but there was potential for increased moisture conservation when fallow land was well managed. This included thorough and deep cultivation of the fallow, good weed control and pest control, a nitrogen status able to make use of the increased moisture, and good seed-bed preparation (ICARDA 1982). Jones (1985) suggests that the best application of fallows might be a system of land management in which sequences of short and long-cycle crops and intervening bare fallows would be planned to optimize water use, since a full profile of stored moisture at planting time permits a crop to produce some yield even in the driest of years. Boersma and Jackson (1977) report the long-practised successful use of summer fallows in semi-arid North America, and point out that a soil depth of one metre is necessary, and preferably 1.5 metres. On the other hand, trials in Israel by Rawitz et al. (1983) showed that the traditional tillage system of deep ploughing and further cultivations in autumn resulted in accelerated erosion and loss of up to 60 percent of the winter rainfall. Basin tillage was more effective, as discussed in Section 5.2.2. Reviewing the result of trials of fallowing in Francophone North Africa, Manichon (1983) concludes that the required conditions for it to work seldom apply in practice. Clearly this is a potentially useful technique but it must be tested in local conditions.

Mixed cropping and interplanting are widely applied traditional techniques. A combination of crops with different planting times and different length of growth periods spreads the labour requirement of planting and of harvesting, and also allows mid-season change of plan according to the rain in the early part of the season (Swaminathan 1982). Another possible advantage may arise from the use of legumes to improve the nitrogen status for the cereal crop. Variations on the theme of mixed cropping, intercropping, and relay cropping are being investigated in the Farming Systems Programme at ICRISAT (1986).

Surface mulching has the advantage of providing protective cover at a time when crop cover is not practical. It improves infiltration, and may also beneficially reduce soil temperature. Possible dis- advantages are:

- the amount of crop residue required may be more than is available from low-level production;
- problems of pest, disease, or nitrogen lock-up;
- the lack of implements which can plant or drill through the mulch;
- organic mulches are liable to be rapidly oxidized in high temperatures.

Successful use of mulching in the semi-arid south west of the USA is reported by Stuart et al. (1985). Trials of different materials and amounts are reported from India (Yadav 1974), and from the dry savanna of northern Ghana (Bonsu 1985).

Timeliness of farming operations is always important, particularly where the rainfall is erratic, and yields can be dramatically affected by planting or cultivating at the right time. Common problems are having to wait for rain to soften the ground because it is too hard to plough when dry, and perhaps then not being able to plant because the ground is too wet. Or a family with only one ox having to wait to borrow another one - hence the interest in the one-ox plough shown in Plate 4.4. Or having to wait for a month after the rains start to get the oxen back into condition for ploughing after a hard dry season. The essence of Farming Systems Research is to look at the whole farming operation to identify the constraints or bottle-necks before starting component research on parts of the system.

Some other techniques should be mentioned, but are beyond the scope of this book, so references are given for the interested reader.

- Deep planting of varieties which can germinate from 15 cm deep, and so delay germination until good rains have fallen (New Mexico, Billy 1981), alternatively, soaking seed before planting when it is desirable to accelerate germination.

Plate 4.4 The single-ox harness and hitch being developed by ILCA

· Dry seeding where the onset of rains can be predicted (India, Virmani 1979).

· Improved ox-drawn implements (Ethiopia, ILCA 1985. Kenya, Muchiri and Gichuki 1983).

· Recent developments in tractor-drawn machinery (Australia, Charman 1985).

· Tillage systems (USA, Wittmuss and Yazar 1981; world review, Unger 1984).

4.2.4 Improved Water Use Efficiency

The selection and testing of alternative crop varieties and, the selection and breeding of cultivars for semi-arid conditions is relatively new but shows promise (Oertli 1983). However, Jones (1985) warns that this solution will be neither easy nor simple because the main requirement is the ability to survive drought periods and start growing again when the drought is broken. This is controlled by a complex of little-understood attributes.

Other desirable characteristics are a short growing season, drought resistance, and drought avoidance. The latter means the ability of the plant to adjust its growth habit according to the available moisture, for example, by tillering when moisture is available or going dormant when moisture is short, or only carrying through to ripening a proportion of the seed heads available.

The uncertainty of crop production reduces the opportunity for the effective use of manures and mineral fertilizers. There are possibilities for economic returns for a small investment, for example "many semi-arid soils have a low sorption capacity for phosphate, which means that small additions are sufficient to give a substantial crop response and will usually have some residual effect for several years after. This is just as well, for little sustained increase in productivity will be possible in these areas without an improvement in phosphate ability." (Jones 1985). There is also evidence that the availability of potassium can improve water utilization through its effect on turgor pressure or the mechanism of stomatal regulation (Lindhauer 1983).

Supplementary irrigation can be important because the provision of small quantities of water at critical times can have good results, for example to allow earlier planting, life-saving irrigation to carry crops through dry periods, or to increase the availability of soluble plant nutrients.

4.3 MECHANICAL CONSERVATION WORKS

4.3.1 Principles

There are no universal conservation practices that work everywhere. Planning soil conservation is like having a large array of techniques and practices set out each in a separate pigeonhole. The object of planning soil conservation is to make up a system by selecting a set of individual items which are each relevant to the conditions, and which can be combined into a workable system.

Looking at the large choice of mechanical works, the main factor in deciding which to select must be to define the objective. The way that different terraces will help meet different objectives is set out in Table 4.1.

The main objective may be:

- to modify the soil slope (Types I, 2, and 3);
- to influence the surface run-off (Types 4 to 7);
- to allow the agricultural use of steep slopes (Type 8).

TABLE 4.1

TERRACES FOR DIFFERENT OBJECTIVES

Objective	Type of terrace
	1.Level terraces for irrigation Plate 4.5
Soil Management	**2. Bench terraces built in a single operation Figure 4.1 and Plates 4.6 and 4.7**
	3.Progressive reduction of slope (fanya juu) Figure 4.2 and Plates 4.8 and 4.9
	4.Absorb all rain (murundum), Plate 4.10
Water Management	**5. Absorb some rain with emergency overflow (contour bund) Figure 4.3 and Plate 4.11**
	6.Controlled run-off (graded channel terrace),Plate 4.12
	7.Controlled reduced run-off - ridging, Plate 4.13 - tied ridging, Plate 4.14
Crop Management	**8.Intermittent terraces, Figure 4.4** - orchard terrace, Plate 4.15 - platforms - hillside ditches, Plate 4.16

Plate 4.5 Level irrigation terraces in Bhutan

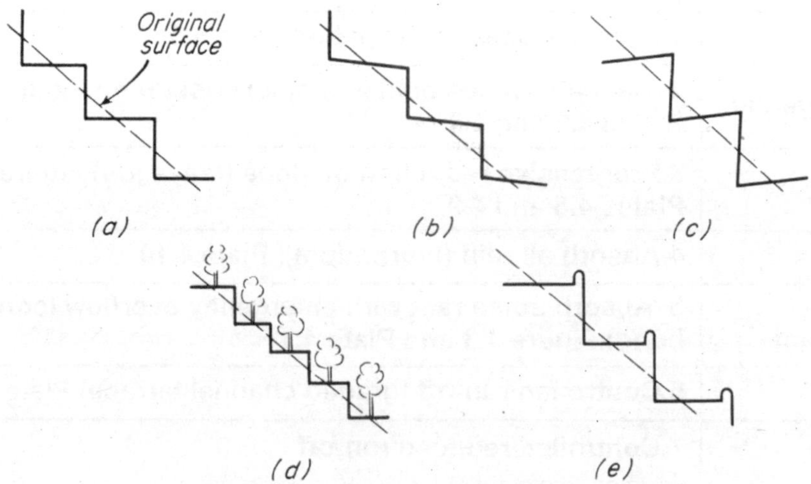

Figure 4.1 Types of bench terraces
(a) Level bench (Plate 4.6)
(b) Outward-sloping Bench
(c) Inward-sloping (or reverse slope) bench
(d) Step terraces (Plate 4.7)
(e) Irrigation terraces (Plate 4.5)

Plate 4.6 Bench terraces in the Yemen Arab Republic (H. Vogel)

Plate 4.7 Step terraces

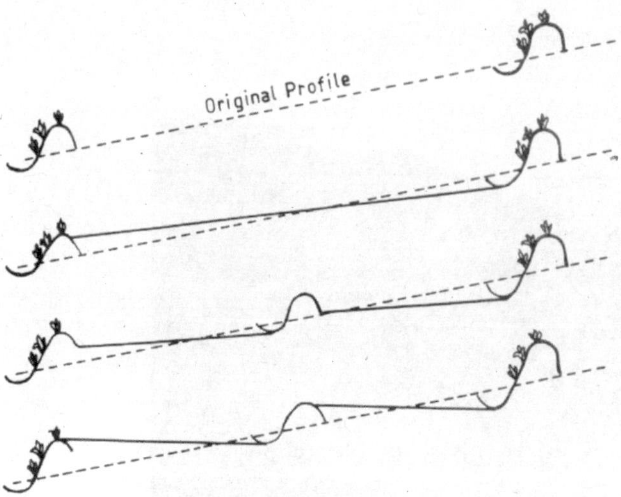

Figure 4.2
Progressive stages in the development of level terraces, using the "fanya juu" system

1. Terraces built at 2 m vertical interval by trhrowing excavated soil up-hill to form a ridge, which is planted with fodder grasses.
2. Movement downslope by cultivation and erosion starts to build up a lower terrace.
3. Main terrace banks built up higher, and intermediate terrace added.
   Further levelling occurs.
4. Final profile is nearly level terraces with well- vegetated tarrace risers.

Plate 4.8
Fanya juu terraces in Kenya
(D.J. Thomas)

Plate 4.9
Vegetation planted on a fanya juu terracve
(D.J. Thomas)

Plate 4.10 Murundum terrace in Brazil, intended to hold all surface run-off, even during heavy storms (T.F. Shaxson)

Plate 4.11 Contour bunds are intended to catch both soil and water

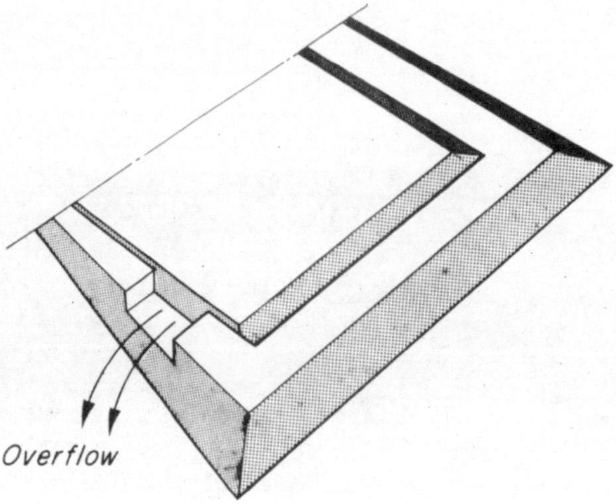

Figure 4.3 Sketch of a contour bund to retain both soil and water (Plate 4.11)

Plate 4.12 Graded channel terraces

Plate 4.13 Ridging

Plate 4.14
Tied ridging in a field in Tanzania (Ministry of Agriculture, Forestry and Wildlife, Tanzania)

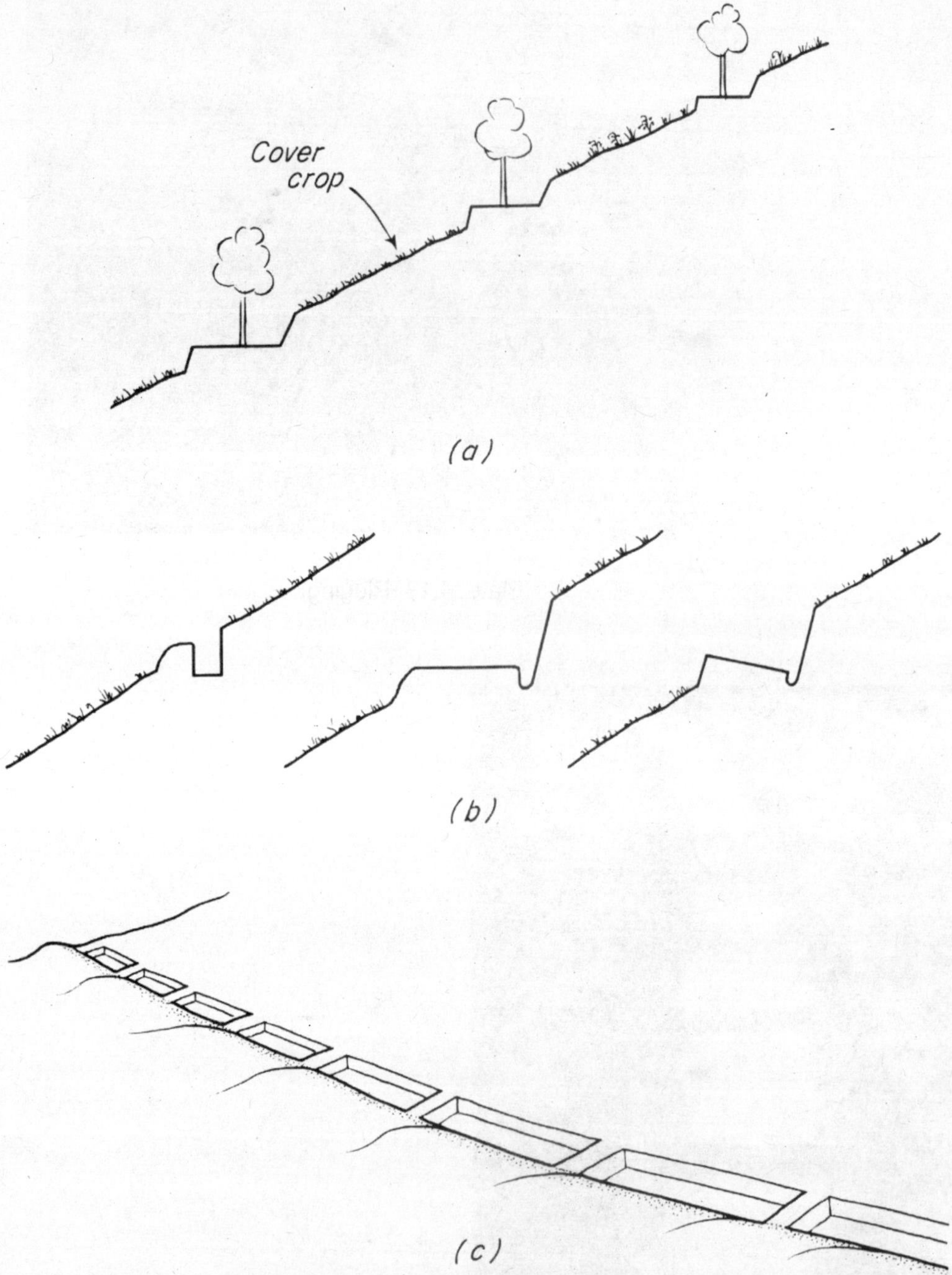

Figure 4.4 Intermittent terraces
(a) Orchard terrace (Plate 4.15)
(b) Types of hillside ditches
(c) Lock and spill drain (Plate 4.16)

Plate 4.15 An example of orchard terracing planted to rubber in southern Thailand

Plate 4.16
Hillside ditch with lock and spill drain (USDA)

In high rainfall areas a common objective is to lead unavoidable surface run-off safely off the land using drains and ditches. In semi-arid regions the objective is more likely to be to slow down the run-off to non- scouring velocities and to encourage infiltration or deposition of silt, without diverting the run-off. This requires simple low-cost structures quite different from the classical system of diversion drains, graded channel terraces, and disposal waterways. That is a high-technology layout of carefully designed structures, and the design procedures are set out in Hudson (1981). The approach is not suitable for semi-arid regions where it is unlikely that there will be suitably trained staff. Simpler techniques are required which can be laid out by village extension workers, or the farmers themselves.

In developed countries a big soil conservation issue is whether the result justifies the cost. In the semi-arid areas this is complicated by the limited alternatives. A dispassionate scientific appraisal may say that some degraded land is best abandoned rather than trying to reclaim it with expensive soil conservation works, but if no better land is available for the production of needed food, then high-labour inputs may be acceptable as the only available option.

There are several well-tested methods for laying out lines either on a level contour or on a predetermined gradient. The A-frame has been widely and successfully used in Africa and in South America, and so has the water tube. In Kenya the line level is preferred. These and other simple levell- ing devices have been compared by Collett and Boyd (1977). Where large areas of gently sloping land are to be laid out, a simple pendulum device can be mounted on a tractor and this has been successfully used in Northern Territory of Australia (Fitzgerald 1977). Whatever method is used to lay out the lines, it is a good idea to make a permanent mark if a tractor or oxen are available. The temporary markers used when laying out the lines are easily lost or disturbed if there is a delay between surveying and construction. Also if channels or earth banks are going to be made by hand, the labour requirement can be reduced by ripping or ploughing by tractor or animals.

Any system of lines, banks, or bunds on the contour has the import- ant by-product of encouraging cultivation on the contour. This alone can result in a reduction of run-off and soil loss of up to 50 percent.

4.3.2 Terracing

Of the types of terrace shown in Table 4.1, few are likely to have widespread application in semi-arid areas. Level terraces may be appropri- ate where irrigation is available (Type 1), or intermittent level terraces (Type 8) used for run-off farming as described in Section 5.2.3. Fanya juu terraces (Type 3) offer a way of achieving level terraces by limited input of labour over a period of time (Figure 4.2). Contour bunds may be useful because of the dual purpose of conserving both soil and water,(Figure 4.3) and Plate 4.11.

There may also be circumstances where a combination of shallow soils with limited storage capacity, and heavy rain, results in frequent surface run-off which calls for a system of graded channel terraces, either without storage (Type 6) or with some storage and a designed overflow. The problem is that any such system is likely to be expensive in relation to the productivity of the land, and it is difficult to maintain grassed waterways as disposal channels when rainfall is limited and unreliable.

Level terraces for dryland farming (Type 2) have been extensively used in the past, for example Ethiopia (Plate 4.17), the Yemen Arab Republic, and in the Maghreb countries of North Africa (Algeria, Morocco, and Tunisia). Most were built in the past and nowadays are increasingly not maintained or abandoned as the maintenance becomes uneconomic or impossible because of labour shortages. One example is the Haraz mountains of the Yemen Arab Republic in the district of Manakhah (Plate 4.18). Until recently Haraz has been one of the most densely populated high-mountain regions in the world, with virtually all slopes being terraced or used as rainwater collection areas. Mainly since the end of 1970, large areas of this man-made ecosystem have been abandoned. In this district it is esti- mated that 800 000 males have migrated from the Yemen Arab Republic to jobs in the nearby oil states out of a population of between 5 and 7 million. Several similar examples are recorded on the north coast of Africa where the migration has been across the Mediterranean to Europe.

4.3.3 Water Disposal

We have seen that in semi-arid conditions it is seldom appropriate to divert surface run-off from arable lands, and the same arguments largely apply to cut-offs or diversion drains put in at the top edge of arable land in order to protect it from surface run-off from uncultivated higher land. There could be special circumstances, such as a shallow saturated soil which would be less damaged if water coming down from above could be diverted. The difficulty is that the drain may also divert run-off during gentle storms which might have been usefully absorbed by the arable land.

The use of diversions will therefore be limited to cases where there is uncontrolled flood run-off in a channel or gully which will be wasted unless it is diverted to some useful purpose. This is discussed under run- off farming in Chapter 5. When there is

Plate 4.17 Long-established terraces in Konso, Ethiopia (H. Hurni)

Plate 4.18
Old terraces in the Manakhah, Yemen Arab Republic, now Inadequately maintained (H. Vogel)

a risk that any structure intended to gather run-off may be overtopped in heavy storms, it is important to deliberately provide planned overspills which can act as safety valves and make sure that the run-off is discharged in places where it will do least damage.

4.3.4 Low-cost Measures

The discussion of terracing and conventional conservation works clearly points to the use of simple and easily applied measures. The first of these should always be farming on the contour. This alone can reduce soil loss to approximately half of what it would be with cultivation up and down the slope. We have already seen that although rainfall in semi-arid areas will be less in total, it can still include very damaging storms, and so it will usually be beneficial to have some form of structure which will slow down the surface run-off, encourage the deposition of suspended mater- ial, and reduce the concentration of surface run-off in minor depressions.

Structures on the contour are simpler and cheaper than graded channel terraces for three reasons. First there is no need to set them out on a precise gradient. They should be more or less on a level contour, but small errors are not as important as in the case of graded channel terra- ces. Secondly, where water is to be led off the land, then the spacing between the terraces has to be calculated, because each channel terrace has to handle the water from a given area. There is no point in using the design formulas when structures are either on a level contour, or are not intended to discharge run-off. If the object of structures on the contour is to store the total run-off then they must be designed to do this, as in the case with fanya juu terraces in Kenya (Thomas et al. 1980), or murun- dums in Brazil, discussed in Section 4.3.2. If the structures are perme- able or can be overtopped safely in heavy storms, then the distance between them is immaterial. Thirdly, since there is no attempt to lead water along the structure, there is no problem of trying to handle the discharge in drains or waterways. However, care is needed to avoid the danger of one level contour bank overtopping, and causing a progressive failure of all the lower banks, with the risk of starting gullies. Plate 4.19 shows such a case in Tanzania.

A general term for simple structures on the contour is 'stop-wash lines' which correctly defines their purpose. The form of such lines will depend on what materials are available. On stony ground, using the stones to build rock lines serves the dual purpose of clearing them from the field as well as building the stop-wash lines.

Plate 4.19
An example of damage caused by the failure of a terrace system. overtopping of one contour bank at the top of the field causes the others to fail progressively (Aerofilms Ltd.)

Plate 4.20 Simple stop-wash lines of straw, stones and soil in Konso, Ethiopia (H. Hurni)

Where stones are not available, lines can be formed by piling up crop residues, perhaps with a few shovels of soil, and progressively built up later by adding weeds from hand hoeing. An example is shown from Ethiopia in Plate 4.20. No design is necessary, but the general principle is that there is not much point in building large or high structures, particularly if built from stone, since they will be very permeable, and in general a larger number of small barriers will be more effective than a small number of large structures.

Grass strips can also be used as stop-wash lines, and this was the basis of a national conservation programme in Swaziland. In the 1940s the king issued a royal edict that strips of the indigenous grass were to be left on all ploughed land, 2 m wide at 2 m vertical interval. The rule was rigorously enforced and almost all arable land has grass strips today as shown in Plate 4.21. For lack of sufficient field advisors many of the strips were off-grade, and others were on land which is too steep for erosion to be halted by this method, as shown in Photo 4.22, but erosion in Swaziland would be very much worse if these strips had not been left. In Kenya live hedges are sometimes planted for the same purpose, often sisal, euphorbia, or other drought-resistant species (Photos 4.23 and 4.24). In areas with higher rainfall, grass may be densely planted to cut for fodder and cause a terracing effect (Photo 4.25).

When stop-wash lines are intended to divert water out of small channels, it is desirable to reduce the permeability at this point. This is done using the principle of the reverse filter. The main structure is composed of large stones, then on the upstream side smaller stones are packed, but large enough so that they cannot be washed through the gaps in the large stones. Upstream of the small stones a layer of gravel is added. Water will still flow through the structure, but slowly, and it will build up in the depression and flow out on either side eventually finding a way through the rock barrier and continuing its path down the slope. This same principle can be used on a larger scale for gully control structures.

Some applications of stone lines have the primary objective of water harvesting rather than soil conservation. Run-off from uncropped land hig- her up the slope runs down onto the cropland, and is spread by the perme- able stone lines along with the run-off which starts on the cropland. When this is the objective there will not be a diversion drain at the upper edge of the cropland, and the stone lines should not use the reverse filter. Where the object is to trap and hold sediment behind the stone bunds, and reduce the slope by developing terraces, the reverse filter effect is desirable along the whole length of the bunds if stones of different sizes are available.

This demonstrates the principle that it is always important to be quite clear about the desired objective. Even a simple device like stone lines can be built to help them to remain permeable, or to silt up as quickly as possible, or to silt up in the depressions only -according to the objective.

There are many examples of inappropriate and unsuccessful attempts to use graded channel terraces in semi-arid conditions (Heusch 1985; Roose and Piot 1984). There are also a number of examples of the successful use of small low-cost structures. An example is the Mossi plateau in Burkina Faso where the recommended solution is to build frequent low barriers (20-40 cm high) at 10-25 m spacing, built of a basic structure of laterite blocks and stabilized with grass (Roose and Piot 1984). Another project in Burkina Faso used similar stone lines as illustrated in Plate 4.26 and described by Wright (1984), and the same approach was used successfully in Mali (Hallam et al. 1985; Hallam and Roose 1985). Plate 4.27 shows the effect on the vegetation of the moisture near a simple line of stones. Plate 4.28 shows another application in the semi-arid south-east of Kenya, on an eroded cattle track, and Plate 4.29 a simple stone barrier across a small wash in Mali.

Plate 4.21 Grass strips have been in place on most arable land in Swaziland for the last forty years

Plate 4.22 Grass strips reduce but do not entirely prevent erosion on steep land, Swaziland

Plate 4.23 Live hedges of drought-resisting species such as sisal and euphorbia are used to contain livestock in Kenya

Plate 4.24
A newly planted barrier of Euphorbia balsinifera and Andropogon gayanus planted to slow down run - off in southern Mali (G. Hallam)

Plate 4.25
Grass strips are cut for stall feeding to stock and help terraces to develop (left) in Nakuru, Kenya (C. Matzon) (below) in north east Brazil (H. Lal)

Plate 4.26 Stone lines set out on the contour near Onahigoua, Burkina Faso (OXFAM)

Plate 4.27
Increased moisture and deposited soil improve the grass growth near stone lines, Mali (G. Hallam)

Plate 4.28 Stone lines on eroded cattle track, Nakuru, Kenya

Plate 4.29 A stone barrier across a small wash, Mali (G. Hallam)

5. WATER CONSERVATION

This chapter deals with methods to increase the amount of water stored in the soil profile by trapping or holding rain where it falls, or where there is some small movement as surface run-off. Schemes which are more concerned with catching and storing run-off for use later or at a distance from the storage are discussed in Chapter 6. The division of methods into these two types is arbitrary, so cross-reference signposts have been used liberally. There are also links with the efficiency of water use discussed in Section 4.2 on Biological Soil Conservation.

Some writers have used the terms internal and external catchments. Internal catchments are where the run-off producing area is within the cropped area and are discussed in Chapter 5; external catchments are run-off areas outside the cropped area and are discussed in Chapter 6.

5.1 PRINCIPLES

5.1.1 Choice of Method

There is no simple way of classifying methods of water conservation. One suggestion is to do it by comparing rainfall with crop requirements (Narayana and Ram Babu 1985), giving three conditions:

i. Where precipitation is less than crop requirements; here the strategy includes land treatment to increase run-off onto cropped areas, fallowing for water conservation, and the use of drought- tolerant crops with suitable management practices.

ii. Where precipitation is equal to crop requirements; here the strategy is local conservation of precipitation, maximizing storage within the soil profile, and storage of excess run-off for subsequent use.

iii. Where precipitation is in excess of crop requirements; in this case the strategies are to reduce rainfall erosion, to drain surplus run-off and store it for subsequent use.

The weakness of this approach is that the main feature of rainfall in semi-arid regions is that it is very erratic and completely unpredic- table. There can be wide variations of moisture shortage and surplus, both within and between seasons. A drought year whose total rain is well below the long-term average may still include periods of excessive rain and flooding, while a high rainfall season may include periods of drought.

This makes the choice of method difficult, because the desired objective may change from one season to another. In a dry area it may be sensible to increase surface storage to improve crop yield in most years, but in a wet year this could cause waterlogging and reduce the yield. On the other hand, a drainage system may have the objective of increasing the run-off but also have the undesired effect of exaggerating the effect of a drought. It is therefore not practical either to classify methods according to average conditions, or to design strategies based on averages. The art or science of water management is to reduce the problems caused by non- average events of flood and drought.

Sometimes it may be passible to have dual purpose methods which can be changed mid-season, for example by opening up the ends of contour bunds to shed surplus water after a wet start to the season, or to block outlets for the opposite effect. But not many methods allow this flexibility, and we showed in Section 2.3.1 that little progress has been made in developing methods of predicting the later part of the season on the basis of the earlier part.

In addition to the vagaries of rainfall, there are a host of other variables, the soil, the land use, the farming system, and the social patterns. Throughout the semi-arid regions of the world there is a wide range of indigenous systems and methods, but experience has shown that they do not always transfer well from one set of conditions to another. In a good discussion of this subject, Pacey (1985 Chapter 6) points out that apart from the technical differences, methods must also be compatible with local life styles, social systems, and patterns of administration. Transferring what appear to be simple techniques requires not only the dissemination of information but also the adaptation to local conditions. In this Bulletin, when presenting information on available methods, we have not tried to prescribe a detailed treatment for a particular problem, but rather to set up a large array of possible methods, from which the person on the spot can choose methods to test locally.

When choosing a method or technique we must be aware of the physical and personal difficulties. Because of the variable rainfall we must expect a low success rate. The odds will be slightly better for grass, trees, and shrubs than for grain crops which are more demanding in their moisture requirements. They need rain at the beginning of the growing season, then a number of storms big enough to generate run-off and at reasonably regular intervals, since run-off events too close together will be wasted if they exceed the storage. The soil must therefore have sufficient soil moisture storage capacity to keep the crop growing between run-off events. There will often be difficulties in keeping stock away from improved areas. Wire fencing is expensive, live hedges need time and care to establish them, and thorn hedges need regular maintenance. Maintenance is also required on any structural works like banks or furrows.

Personal constraints are that people naturally do not like to put a lot of effort into schemes which have a low success rate, nor if their tenure is insecure. In dry areas the people are frequently mobile or partly nomadic and so may not live permanently near sites suitable for run-off schemes. There are many former schemes which are now abandoned, and knowing more about the reasons would help us plan new schemes.

5.1.2 Some Design Principles

The term "rainfall multipliers" describes methods where the run-off from an uncultivated part of the land is diverted on to a cultivated part, thus giving it the benefit of more water than it receives directly as rain, hence rainfall multiplier.

The object of designing such methods is to get the best ratio of area yielding run-off to the area receiving run-on. Because of the variations of soil and crop requirement, and above all the variability of rainfall, this can never be a straightforward mathematical calculation. A formula which can be used to provide a starting point for field trials is given by Finkel (1985), together with some empirical values for Turkana in Kenya, to be used in the formula:

Ratio =	Catchment area =	Crop water requirement - Design rainfall
	Cultivated area	Design rainfal x Run-off Coef' x Efficiency

· Crop water requirement is estimated from evaporation data with a crop factor applied;

· Design rainfall is probably less than long-term average rainfall in order to give a better crop in low rainfall years, while the excess can overflow in wet years;

· Run-off coefficient is the percentage of rainfall which becomes run-off;

· Efficiency reflects the difference between rainfall distribution and the water requirements of the crop.

Schemes with internal catchments require high energy inputs for surface modification and so tend to be more common on crop land and in areas with annual rainfall of 250 mm or more. The ratio of catchment area to cultivated area is likely to be from 1:1 to 5:1. Schemes using external catchments may have much higher ratios, for example up to 30:1 in the case of the run-off farming in the Negev desert (Evenari et al. 1982). Such systems may be practical at levels of rainfall down to less than 100 mm, which is lower than is possible for in-field schemes for cropland.

Two examples of rainfall multiplier:

a) Internal catchment, ratio 2:1, design rainfall 250 mm

2	= 500 - 250	= 250
1	250 x 60% x 80%	120

b) External catchment, ratio 20:1, design rainfall 100 mm

20	= 500 - 100	= 400
1	100 x 25% x 80%	20

The run-on is greater in b), so the soil moisture storage must be greater, or run-off events better distributed.

In developed economies it is reasonable to look at the long-term returns from schemes which require investment, but this is less helpful in semi-arid areas. Even when long-term rainfall records are available, and probabilities can be calculated, subsistence farmers or nomadic pastora- lists work to very short-term plans. Several examples are reported from Kenya by Finkel (1984) when soundly planned schemes were abandoned after poor results in the first year to two.

5.1.3 The Effect of Scale

The size of catchment areas has a bearing on the yield of run-off. Under the same hydrological conditions, small areas may have up to 50 percent, whereas river basin run-off is only 5 percent (Stern 1979). There are two reasons for this. Large catchment areas are more likely to have places where there is temporary storage of run-off in surface depressions. Also small catchment areas may yield run-off from short showers which do not produce surface run-off from larger areas. The hydrological explanation is that small areas have a shorter gathering time (i.e. time of concentra- tion) and so give run-off from shorter storms (with higher intensities) than storms which need a longer gathering time to produce run-off from larger catchment areas. Shanan and Tadmor (1979) quote the following average run-off yield from 100 mm annual rainfall:

Less than	0.02 ha	10-30 mm/ha/year
	3-10 ha	4-10 " " "
	300-500 ha	1 " " "

Procedures for estimating rate and amount of run-off are given in Hudson (1981) and FAO (1981).

5.2 METHODS FOR CROP LAND

The techniques described in this section all involve some re-shaping of the soil surface, and so require substantial inputs of energy. In many cases it is not practical to use only hand labour, and animal-drawn imple- ments are necessary, or tractors. Some of the methods show promise of increased crop yields so a major feature of experimental trials is the search for cheap, simple, low energy methods and machines. Figure 5.1 shows the main features of the methods described in this Section.

5.2.1 Broad Bed and Furrow System (BBF)

The Broad Bed and Furrow system has been mainly developed at the International Crops Research Institute for the Semi-arid Tropics (ICRISAT) in India (Krantz 1981, Pathak et al. 1985). A comprehensive research programme was carried out on-station for eight years before being taken for on-farm adaptive research at Tadthanapalle in Medac district (ICRISAT Annual Reports 1975-1984). It is a modern version of the very old concept of encouraging controlled surface drainage by forming the soil surface into beds. In medieval times in Britain this was used for improving pastures and called "rigg and furrow"; it has also been used in North America (Phillips 1963) and in Central Africa (Cormack 1954). A variation known as the camber-bed system was used in Kenya (Robinson et al. 1955). An experimental layout at ICRISAT is shown in Plate 5.1 and the bed cross-section in Plate 5.2.

The recommended ICRISAT system consists of broad beds about 100 cm wide separated by sunken furrows about 50 cm wide. The preferred slope along the furrow is between 0.4 and 0.8 percent on vertisols. Two, three, or four rows of crop can be grown on the broad bed, and the bed width and crop geometry can be varied to suit the cultivation and planting equipment. Some examples are shown in Figure 5.2.

In India the system has been used mainly on deep vertisols (heavy black clay soils sometimes called cotton soils); wide beds are used on a gentle grade and they are formed by ox-drawn wheeled tool carriers (Plate 5.3). There are extensions of this work on similar soils in Ethiopia (ILCA 1985) with an interesting development of a very simple ridging implement shown in Plate 5.4. Another variation using smaller ridges is being developed by the Agricultural Research Centre for the Semi-arid Tropics (CPATSA) of the Brazilian Enterprise for Agricultural Research (EMBRAPA) located at Petrolina, Brazil (Lal 1985 and 1986); it is illustrated in Plate 5.18.

All these schemes have, in varying proportions, the following objectives:

i. to encourage moisture storage in the soil profile. Deep vertisols may have soil moisture storage up to 250 mm, which is sufficient to support plants through mid-season or late-season spells of drought. The possibility is also increased of double cropping by means of inter-cropping or sequential cropping. The large water storage capacity of the soil supports growth more easily during the subsequent dry but cooler post-rainy season.

ii. to dispose safely of surplus surface run-off without causing erosion. Plate 5.5 shows the flooding which can occur on vertisols.

iii. to provide a better drained and more easily cultivated soil in the beds. There is only a narrow range of moisture conditions during which the soil can be efficiently tilled or planted, and timeliness is a key factor. Only about 20 percent of the deep vertisols in India are cropped during the rains, mainly because of poor workability when wet. The situation is similar in Ethiopia (Plate 5.6). If a crop can be established during the early rains, the profile is usually near saturation only for short periods during the latter half of the season, water is more efficiently utilized, and there is less need for run-off collection and storage. The possibility is also increased of double cropping by means of inter-cropping or sequential cropping (Kampen 1979 and 1981). Tillage of the raised beds may be possible before the rains, introducing the possibility of dry seeding ahead of the rains in areas where the start of the rainy season is fairly reliable, and there is a good chance of follow-up rains to ensure the establishment of the germinating crop. The difficulty of preparing a seed bed during the dry season in these hard clay soils has been greatly improved by the use of broad beds and animal-drawn equipment.

iv. the possibility of the re-use of run-off stored in small tanks (Plate 5.1). Small amounts of life-saving irrigation applications can be very effective in dry

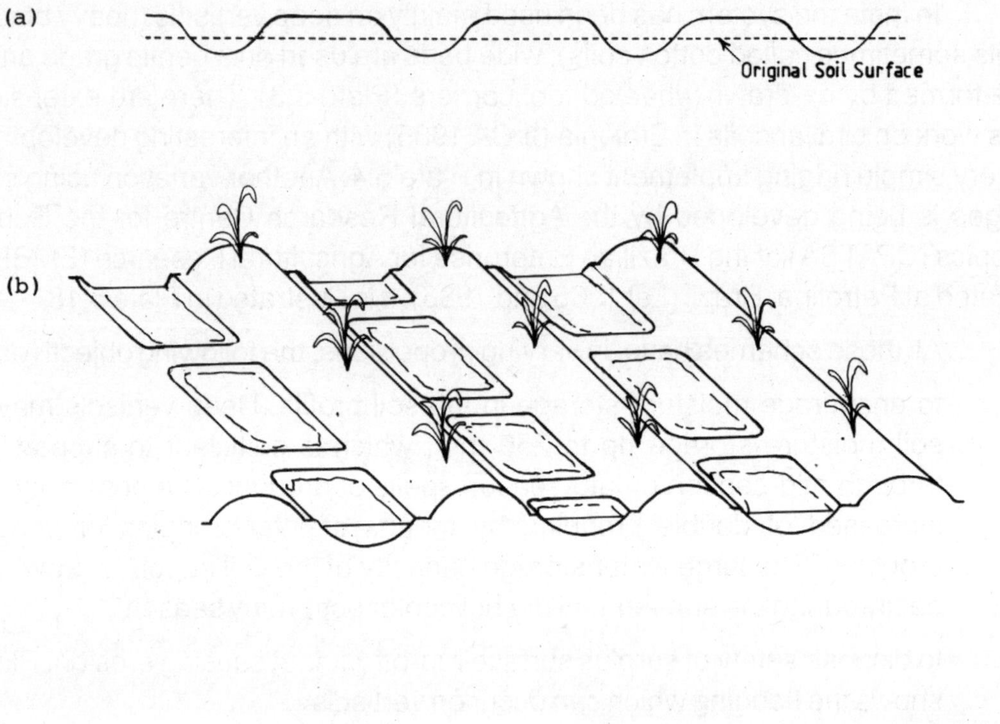

Figure 5.1 Methods of changing the soil surface to increase retention of rainfall on crop land:

(a) Broad bed and furrow
(b) Tied ridges
(c) Conservation bench terraces
(d) Contour furrows or strip tillage

Plate 5.1 An experimental catchment at ICRISAT with broad bed and furrow layout, grass waterway for safe removal of run-off, and storage tank for supplementary irrigation (ICRISAT)

Plate 5.2 Cross - section of broad bed and furrow

Plate 5.3 One of several tool carriers tested at ICRISAT for forming and cultivating beds (ICRISAT)

Plate 5.4 A simple ox-drawn ridger being developed at ILCA in Ethiopia

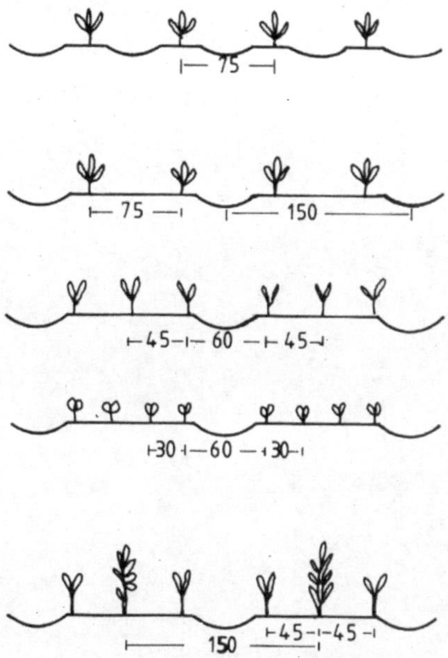

Narrow beds and furrows are adapted to 75 cm rows only (e.g. maize)

Broad beds and furrows are adapted to many row spacings
A maize crop

A sorghum or millet crop

A groundnut or chickpea crop

A pigeonpea/sorghum intercrop or a pigeonpea/maize intercrop

Figure 5.2 Alternative crop and row arrangements on broad beds

Plate 5.5 Run-off and erosion on a vertisol in India

Plate 5.6 On heavy clay soils in Ethiopia, wide beds with open furrows provide drainage

Plate 5.7 Trials at ICRISAT of a 3-row planter on broad beds (ICRISAT)

spells during the rains, particularly on soils with lower storage capacity than the deep vertisols.

The BBF system is particularly suitable for the vertisols. The technique works best on deep black soils in areas with dependable rainfall averaging 750 mm or more. It has not been as productive in areas of less dependable rainfall, or on alfisols or shallower black soils - although in the latter cases more productivity is achieved than with traditional farm- ing methods. Other methods, with more emphasis on storage and irrigation within a package which includes BBF, are more likely to be viable for the alfisols (Ryan et al. 1979). It is also stressed throughout the ICRISAT research that the BBF system should not be considered in isolation, but only as part of an improved farming systems package.

An important component of the system is an ox-drawn wheeled tool bar, which can be used with ridgers to form the raised bed and also later for carrying precision seeders or planters (Plate 5.7). The tool carrier is thus used for the initial forming of the beds, the subsequent annual reshaping, and for all tillage, planting, and inter-row cultivation. Even the cost of the simplest factory made tool bar is beyond the resources of most peasant farmers, but the really cheap wooden ridger shown in Plate 5.4 shows promise.

5.2.2 Ridging and Tied Ridging

This method is also known in the USA as furrow blocking, furrow damming, furrow diking, and basin listing. The principle is to increase surface storage by first making ridges and furrows, then damming the furrows with small mounds, or ties (see Plate 4.14). Graded ridges alone will usually lead to an increase of surface run-off compared with flat planting, while tied ridges will decrease the run-off and increase the storage. In different seasons either of these two effects may be prefer-able. And so the design question is when to go for drainage and when for storage. An aid to designing what is best in the long-term by modelling has been developed (Morin et al. 1984), but designing the system which will give the best effect in the long term will not prevent the system giving the opposite effect from that which is required in a particular season. The possibility has been suggested by Ahn (1977) of hedging the bet by tying alternate furrows. This is not a very satisfactory solution. One argument is that this would reduce the amount of damage by too much run-off or too much retention, but one could equally well argue that it will ensure that half of the land will always be under the wrong treatment.

An interesting possibility is to use tied ridging in connection with sprinkler irrigation to allow higher application rates at low pressure (Stuart et al. 1981). Tied ridging is usually associated with mechanized farming. There have been some attempts at achieving it with ox-drawn implements, but the system really needs high draught for speed and preci- sion, which is required if the ridges are to be re-ridged or split in subsequent years.

Either ridging alone or tied ridging has occasionally been practised using hand labour, but the high labour requirement usually makes this unpopular with subsistence farmers. Anyway hand-made ridges are usually less efficient. They are more likely to depart from a true contour and to have variations in the height of the ridge, both of which will increase the risk of overtopping. A variation in the theme, which is suitable for hand work, has been tried on a limited scale in Burkina Faso and is called tied-mounding shown in Plate 5.8.

Plate 5.8
Experiments with artificial rain - fall on 'tied-mounding', Burkina Faso (J. Collinet).

i. Varying evidence of the results of tied ridging

There is an extensive literature reporting trials of tied ridging in many countries. A few of the reports indicate problems and failures but the great majority claim such outstanding success for the system that one wonders why the system has not been more widely used. A possible solution to this conundrum may be that the inconsistency, and unreliability of good results, prevents it becoming more widely adopted.

The experience in Africa has been well summarized by El-Swaify et al. (1985) as follows:

"Under certain circumstances the system has been beneficial not only for reducing run-off and soil loss, but also for increasing crop yield (Lawes 1961, 1963; El-Swaify 1983; Dagg and McCartney 1968). However, during high rainfall years or in years when relatively long periods within the rainy season are very wet, significantly lower yields were reported from systems with tied ridges than from graded systems which disallowed surface ponding ef water (Lawes 1963; Dagg and McCartney 1968). Under such conditions tied ridging enhanced waterlogging, developed anaerobic conditions in the rooting zone, excessive fertilizer leaching, and water table rise in lower slope areas."

In other countries there are conflicting reports, with a majority of successes. McCartney et al. (1971) reported that tied ridging in Tanzania gave higher maize yields not only in low but in high rainfall years as well. However, reports of success are more common in low rainfall years, for example Njihia (1979) reported from Katumani in Kenya that tied ridging resulted in the production of a crop of maize in low rainfall years when flat-planted crops gave no yield. Jones (1985) goes on to say:

"Similarly on a sandy soil at Lusitu in the Zambesi valley, tied ridges increased mean crop yields (maize, sorghum, and millet) over those on flat land by 168, 159 and 16 percent under seasonal rain- falls of 587, 623, and 724 mm (Honish, 1973); and on vertisols at Big Bend, Swaziland, mean increases for maize, cotton, and sorghum were 64 percent in a year of 508 mm, and 308 percent in a year of 310 mm (Warwick 1979, 1980). Clearly responses can be dramatic, but recent work under very harsh conditions in Botswana has shown that there may be also negative effects. Higher soil temperatures within the ridge can be detrimental to seed germination, and where showers are light the penetration of moisture into the soil may be shallower than that in the flat soil (DLFRS, 1984)."

Another report from Botswana indicates a positive response to sorghum yields from tied ridging in both a very dry season (1972/73) and a very much wetter one (1973/74) (DLFRS, 1973; 1974).

In the Negev desert in Israel, Rawitz et al. (1983) reported that tied ridging greatly increased the effi- ciency of a fallow before cotton, but that when the system was tried for wheat crops it was less successful because of waterlogging in the basins (Plate 5.9). The use of wide beds with untied furrows was therefore pre- ferred and gave increased yields, but the climate of the Negev is special since it has winter rainfall and a very low mean annual rainfall.

Plate 5.9 Basin listing for wheat production in Israel (D. Rosenzweig)

From India also, broad ridges gave better yields of sorghum than either flat planting or narrow ridges(ICRISAT 1976) and a similar result was reported for sorghum and castor in Gujarat by Brahmbatt and Patel (1983). In Texas, USA, Clark and Jones (1981) reported substantial increases in sorghum yield from 1420 kg/ha to 1650 kg/ha for tied ridging compared with flat planting. In another American study in Texas diked furrows were found to be effective at holding and absorbing high intensity storms (Billbro and Hudspeth 1979). Similar successful applications of tied ridging in the Great Plains of the USA are also reported by Stuart et al. (1985), and again one is inclined to wonder why, in the face of so much positive experimental evidence, one does not find the whole of Texas and the whole of Tanzania covered with tied ridging.

One possible reason for the low adoption of the system is the temperature effect reported by Jones. Another is the danger of waterlogging in wet years. This can be worsened when there is compaction of the furrow bottom by tractor traffic, and one must also consider whether the machinery and implements are as efficient as the reports would suggest.

ii. Conditions where tied ridging is suitable

There is a danger of soil erosion if the ridges are overtopped and break so that the water temporarily stored in the depressions is suddenly released. This will not happen if the combination of surface storage plus the amount which infiltrates into the soil surface is less than the storm rainfall. This implies a high value of soil storage, usually deep soils with good infiltration and permeability. In some systems the infiltration is increased either by mulching in the furrow bottoms or by subsoiling or cultivating.

Three safety back-ups are required to minimize the risk of damage by erosion:

· the furrows should be on a gentle grade to assist run-off if the ties fail;

· the ties should be lower in height than the ridges so that the ties fail along the furrows before the ridges fail down the slope;

· there should be a back-up system of conventional graded channel terraces to prevent damage if the ridges do overtop or fail.

iii. Implements for building tied ridges

Making an implement to form the ridges is straightforward, it is interrupting the ridging process to leave a tie that is difficult. Possibilities are:

· intermittent lifting by hand if the ridger is pulled by tractor or by oxen (Plate 5.10);

· automatic lifting devices based on an eccentric wheel (Plate 5.11);

· intermittent hydraulic lift either manual or triggered by rotation of tractor wheels.

The design of implements in the USA is described by Lyle and Dixon (1977), and in the USSR by Tregubov (1981). An earlier review by Boa is useful but out of print (FAO 1966).

iv. Simpler systems for non-mechanized farming

The construction of tied ridges is mainly associated with mechanized farming because of the high labour requirement, but it is quite possible to do it by hand labour. Plate 5.12 shows an example from Ethiopia where the ridges and ties are made by piling up lines of sorghum residues and roots and soil.

Plate 5.10 Making the ties by lifting the ridger by hand, malawi (T.F. Shaxson)

Plate 5.11 The ridger lifted by eccentric wheel (T.F. Shaxson)

Plate 5.12 Building ridges and ties from crop residues, roots and soil, Ethiopia (H. Hurni)

5.2.3 Conservation Bench Terraces (CBT) (also known as Zingg terrace, and flat channel terrace)

This is another type of rainfall multiplier, using part of the land surface as a catchment to provide additional run-off onto level terraces on which crops are grown. The method is particularly appropriate for large- scale mechanized farming such as the wheat/sorghum farmlands of the south- west of the USA, where the method was pioneered by Austin W. Zingg in 1955. Experimental results were first reported in 1959 (Zingg and Hauser) and after detailed studies at Bushlands Experiment Station in Texas, and Hayes in Kansas, a technical and economic evaluation was made by Hauser and Cox (1962). This led to extensive trials in the six western States with low rainfall, and these were reported in detail (ASAE 1968). CBTs were compared with the conventional practice of level terraces (i.e. level along the length, but the original slope is left between terraces), and also with all over bench terracing (Figure 5.3). There is a soil erosion hazard during high-intensity summer storms, but CBTs were as effective at controlling erosion as the other two practices, and more effective at reducing the overall run-off. The data from these trials provided general guidelines on the method, but standard designs should be avoided because of the wide variation in conditions of soil, rainfall and farming system. The best way of applying the system in a particular situation should always be investi- gated locally.

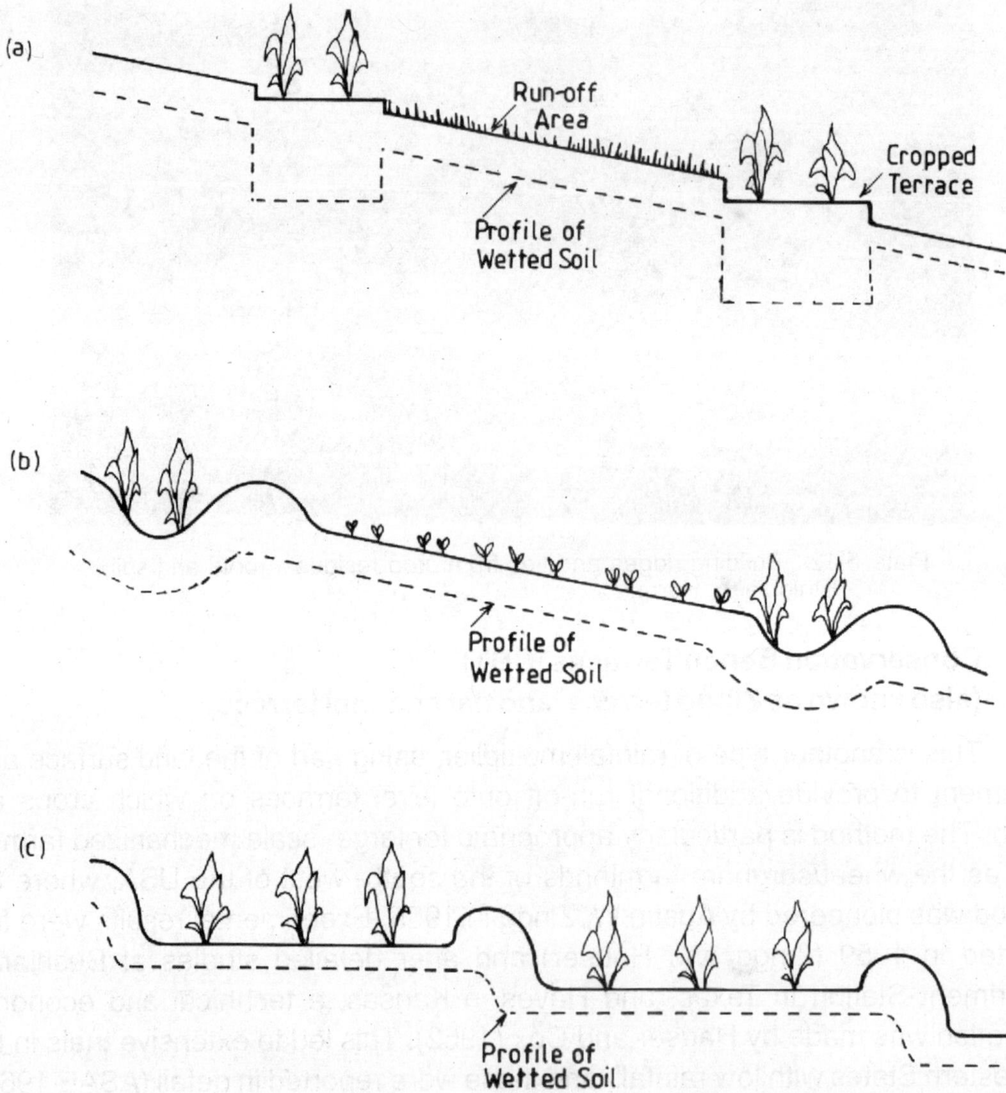

Figure 5.3 Conservation bench terraces compared with other practices:
(a) conservation bench terraces increase soil moisture storage under the terrace
(b) contour bunds also increase the soil moisture under the bund
(c) level bench terraces do not heve multiplier effect

The features required for CBTs are:

· Gentle slopes of 0.5 - 1.5 percent are most suitable although the system has been used up to 6 percent in North Dakota. As with all terracing, a steeper slope requires more earth moving.

· A deep soil is required, both to provide sufficient soil moisture storage, and also to lessen the effect of cutting during the con- struction of the terraces. Several USA workers report a reduction of yield from the disturbance of the soil during terracing for up to six years. Some even went to the length of stockpiling the topsoil and replacing it after levelling. Good permeability is also required so that the contained flood water can be absorbed quickly.

· Smooth slopes are an advantage where large mechanized farming can be made more convenient by constructing all the terraces parallel and of equal width.

· Precise levelling of the bench terrace is important to ensure uniform build up of soil moisture.

· If there is a risk that run-off from the catchment area will be greater than can be absorbed and stored on the terrace, there must be outlets at the ends of the terrace, which can discharge into grass waterways or other safe disposals. As with tied ridging, if there is going to be surface run-off it must be along the terrace, not over the edge of the terrace and down the slope with consequent risk of erosion. For CBTs a common method is to make the level of the outlet at the end of the terrace one half of the height of the bund on the edge of the terrace, ensuring a freeboard of half the bund height.

· This is a high-cost system which has to be paid for by high yields and so a high level of inputs is required, particularly seed, fertilizer, and crop management. Adequate levels of fertilization are required to avoid this being the constraint after the moisture limitation has been removed.

· The main design factors are the width of the levelled terrace and the ratio of the area of the terrace to the catchment area. The width of terrace is governed mainly by the machinery to be used. As the width of planters and drills continues to increase, so does the need for wider terraces, but this is restricted by the slope and the depth of soil. Typical widths are from 10 metres on land of 5 - 6 percent, to 30 metres at 2 percent, and 50 metres or more at one percent. Jones (1981) reports mini-terraces 9 m wide with a 1:1 ratio and different rotations on the catchment and the terrace. Total yield was doubled on the terraces.

· The most usual practice is for the bench terrace to be level along its length, but there have been a few trials with the terraces on a gentle gradient of 1:400 to encourage safe disposal of excessive run-off.

· The ratio of catchment to terrace may be from 0:1 (in which case the whole land surface is converted to level bench terraces) but typical values are 1:1 or 2:1. In general, the lower the mean annual rainfall the larger will be the required catchment area to provide sufficient moisture for the crop. Another consideration will be the type of rainfall since run-off will only be produced from storms where the intensity exceeds the infiltration on thecatchment area. The treatment of the catchment area will also effect the amount of run-off. In very dry areas the catchment may be left uncultivated to promote run-off, while in the south western USA the usual practice is to have continuous cropping on the bench terrace with a rotation on the catchment area. A common rotation on the catchment is wheat/sorghum/fallow, giving two crops in three years. Table 5.1 gives examples of the rainfall multiplier effect.

Results of the American experience show that the method is suitable and economic for the high-input large-scale mechanized farming of the region. The costs of land preparation can be expected to be recovered in about 10 years. Complete bench terracing of the whole land surface some- times gives the greatest cash return since the whole of the land is cropped, but the cost of terracing is much higher. With a ratio of 1 terrace:2 catchment the terracing costs are only one third, and in several studies this was the most economic system (Jones and Shipley 1975).

TABLE 5.1

THE RAINFALL MULTIPLIER EFFECT OF CONSERVATION BENCH TERRACES

Assuming 500 mm of water is required for a sorghum crop*

	Catchment Ratio	Run-off Coef^{t} C%	Run-on (MARxRatioxC)	Total
MAR = 250 mm	2	40	200	450
	2	50	250	500
	3	30	225	475
	3	40	300	550
	4	30	300	550
MAR = 350 mm	2	20	140	490
	2	25	175	525
	3	10	105	455
	3	15	157	507

* The actual amount will depend on the water use efficiency, i.e. allowing for losses from seepage and evaporation.

Note: An example of CBT being used for growing winter wheat comes from Arizona. The average winter rainfall is 180 mm and assuming a moisture requirement of 585 mm and run-off of 33%, a catchment ratio of 6.8 was used (See Section 5.1.2):

$$180 + \frac{(33 \times 180 \times 6.8)}{100} =\sim 585$$

The calculated bench was 2.4 m wide, and the catchment 16.3 m (data from Flug 1981).

These calculations assume 100% efficiency, compared with Finkel's method.

Because of the high cost of installation of CBTs, it is not appro-priate at very low rainfalls. The main application of the system is to increase the yield and reliability of yield where rainfall is nearly sufficient for crop production, in the range 300-600 mm. Improving the probability of getting a reasonable crop may be more important than numerical increase of yield. For methods more suitable in lower rainfalls see Section 5.2.4 for simple methods, and Section 5.3 for methods which use water spreading.

5.2.4 Contour furrows (also known as contour bunds and desert strip farming)

These are variations on the theme of surface manipulation which require less soil movement than conservation bench terracing, and are more likely to be used by small farmers, or in lower rainfall areas. The crop- ping is usually intermittent on strips or in rows, with the catchment area left fallow. The principle is the same as Conservation bench terraces, that is to collect run-off from the catchment to improve soil moisture on the cropped area. Contour bunds as used in Kenya, are shown in Figure 5.4. Studies were made of the soil moisture profile (Smith and Critchley 1983) with a catchment ratio of 2:1; a satisfactory sorghum crop was grown on only 270 mm of rainfall. It was estimated that run-off from the catchment was 30 percent giving 166 mm of run-on, and 432 mm available to the plants:

$$\frac{(30 \times 2 \times 270)}{100} + 270 = 432$$

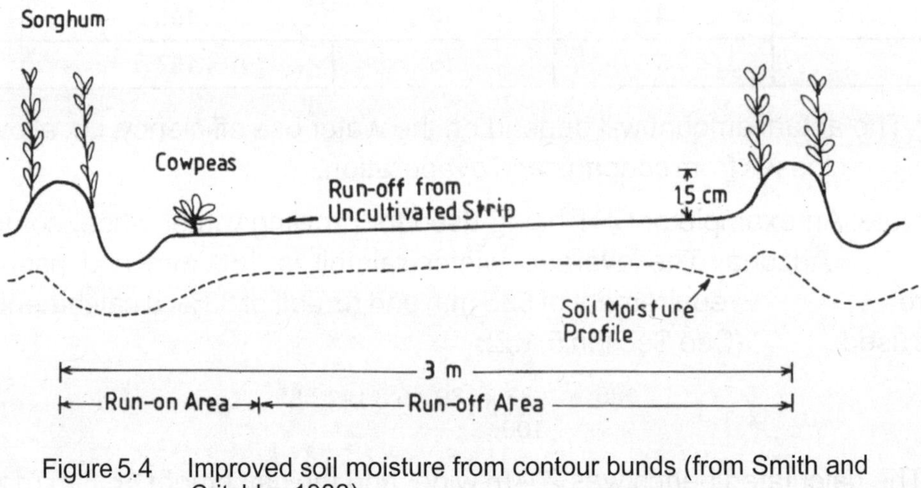

Figure 5.4 Improved soil moisture from contour bunds (from Smith and Critchley 1983)

On heavier soils, contour bunds may be less effective because of the lower infiltration. Studies on vertisols in India showed that yields were lower near the bunds, both upslope and down, as a result of waterlogging (ICRISAT 1976).

Contour bunds are also used in Ethiopia (Plate 5.13) for a combina- tion of soil conservation and water conservation. The bunds are built on level grade with ties in the basin. A stone wall is built on the lower side of the earth bund in an attempt to reduce damage if the basin is over- topped, but this has not been very successful, and it would be better to put the bunds on grade if it is likely that they will not be able to hold all the run-off (Hurni 1984). Contour bunds are also discussed in Section 4.3. Another successful example, from Ethiopia, is shown in Plate 5.14.

If the contour furrows are not laid out precisely on the contour, or are built with some irregularities, there may be a danger of uneven depths of ponding behind the bank. This can be reduced by smaller bunds at right angles, but as with tied ridging, these bunds should be lower in height than the main ridges so that if there should be any overtopping it will be laterally along the contour and not over the bund and down the slope. Sometimes the emphasis is on the excavated furrow which collects water, so that in exceptional storms the run-off can overflow without damage.

A disadvantage of contour bunds at intervals down the slope is that the crop tends to be uneven, reflecting the soil moisture profile in Figure 5.4, and only a small part of the field is cropped. But the point is that it enables some crop to be grown where otherwise the rainfall would be inadequate for any cropping. The energy input for building the ridges is still considerable but several ox-drawn ridgers are available, and more expensive but more efficient tool bars, as discussed in Section 5.2.1. Even if the whole operation has to be done by hand labour this may be acceptable if it is replacing a traditional system which also has a high labour requirement. This was the situation in the Kenya example reported by Smith and Critchley, where the traditional method consisted of deep digging by hand to increase the infiltration and moisture storage.

Another approach to localized surface storage on part of the land is the use of small semi-circular catchments or larger rectangular or trape- zoidal structures (see Figure 5.5). The large number of separate semi- circular or trapezoidal bunds spreads the risk of damage from overflow in exceptionally heavy rain, or if one bund fails. The semi-circles are usually constructed by hand, and are larger versions of range pitting which is used to promote vegetation on grazing land (see Section 7.1 and Plates 7.5 and 7.6). In several semi-arid climates in Africa the farmers have shown more interest in using these techniques for crops than for range improvement (Smith and Critchley, 1983, describing the work in Baringo, Kenya, and Wright, 1984, working in Burkina Faso). The trapezoidal bunds were developed by Finkel (1985)in Turkana, Kenya, and can be built by hand or machinery. Semi-circular hoops built of stones for tree planting are discussed in Section 7.2.2 and illustrated in Plate 7.21.

Plate 5.13 Contour bunds in Ethiopia, with ties in the basin and stone reinforcement of the bank (H. Hurni)

Plate 5.14 Contour bunds in Tigre, Ethiopia, give increased grass growth (J. Bennett, Andes Press Agency)

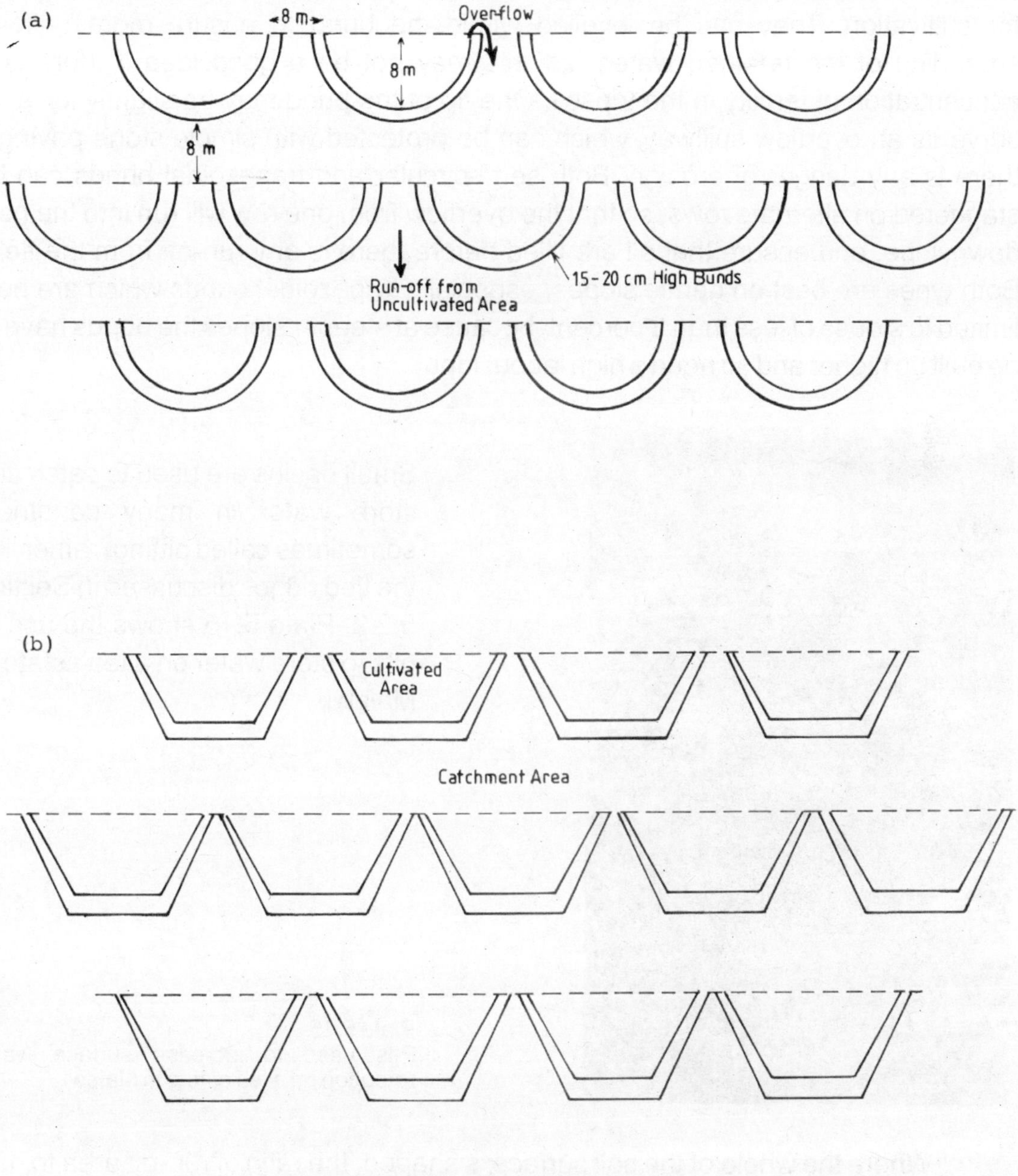

Figure 5.5 Simple structures to retain surface run-off
(a) Semi - circular catchments
(b) Trapezoidal catchpits

The trapezoidal bunds can be from 0.25-2 hectares, and so are more suitable for cultivation. They can be levelled within the bund to ensure more uniform spreading of the retained water, but this may not be a good idea if there is a concentration of fertility in the topsoil. One tip of the bund may be slightly lower to serve as an overflow spillway, which can be protected with simple stone paving if there is any danger of erosion. Both semi-circular and trapezoidal bunds can be staggered on alternate rows, so that the overflow from one row will run into the next downslope, and ensure that all are filled before there is any run-off from the field. Both types are best on gentle slopes, especially trapezoidal bunds which are best limited to slopes of less than 2 percent, because at steeper slopes the bunds have to be built up higher and so need a high labour input.

Small basins are used to catch and store water in many countries, sometimes called pitting, rather like the tied ridges discussed in Section 5.2.2. Plate 5.15 shows the use of pits to store water on a tea estate in Malawi.

Plate 5.15
Pits used to increase surface water retention on a tea estate in Malawi

Where the whole of the soil surface is shaped, the ratio of run-off area to run-on area is governed by the configuration as shown in Figure 5.6 and Plates 5.16 and 5.17. The land may be shaped by ploughing, as in Plate 5.17, or by ridgers used with oxen, as in Plate 5.18, or tractors as in Plate 5.19. An important advantage of small run-off producing areas compared to larger ones is that they will yield run-off from smaller rain storms. This was discussed in Section 5.1.3. At the Central Arid Zone

Research Institute in India, using narrow strips with a ratio of run-off to run-on of 0.75 (catchment) to 1.0 (cropped), threshold values of rainfall for inducing run-off were between 3 and 15 mm for moist surfaces and 7 to 9 mm for dry surfaces. The possibility of increasing run-off through surface applications of bentonite or sediment from tanks was possible but unlikely to be economic. Another possibility which could have practical significance is changing the configuration to increase the slope of the run-off area (Jain 1980). A system of creating furrows on sloping land using asymmetric ridgers was shown to be successful for maize production on an experimental basis in Mexico (Garduno 1981), see Figure 5.7.

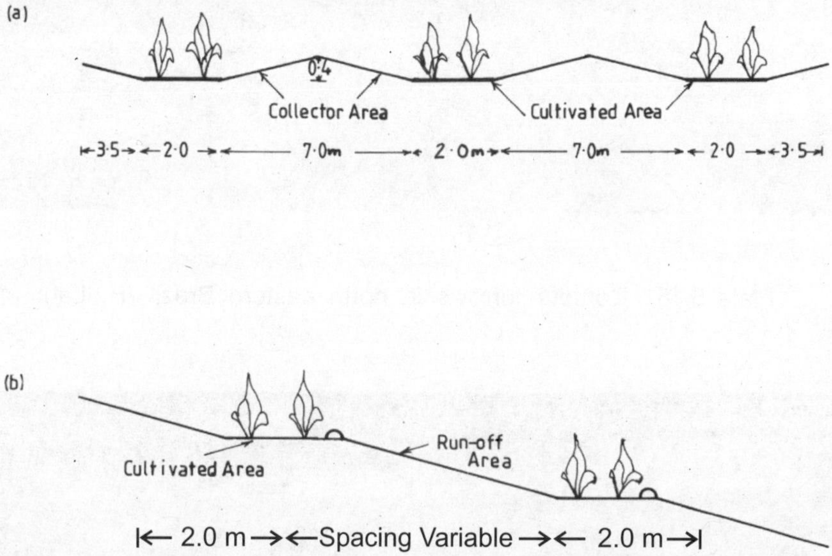

Figure 5.6 Shaping the whole soil surface into run-off and run-on areas: a) on level ground; b) on gently sloping ground

A configuration suitable for flat or very gently sloping land, is shown in Plate 5.20 and comes from studies in Mexico. A similar cross- section but on a small scale was shown to be successful in Colorado by Fairbourn (1975). Low cross-ties can be used to help obtain a better distribution of water since they reduce accidental run-off if some of the lines have a fall. Fairbourn also showed that infiltration could be increased by burying crop residues in a narrow trench in the cultivated area; he called this vertical mulching but the extra complication of trenching and filling make this practice unlikely to be adopted on a wide scale (Figure 5.8). It bears some relation to a simpler variation in which crop residues are thrown into the bottom of furrows and the ridges are split onto the furrows to form a new ridge for cropping the following year. Subsoiling or deep ripping in the furrow may also be used to increase infiltration, particularly when there is a clay pan or a laterite layer.

Plate 5.16 Contour furrows in north eastern Brazil (H. Lal)

Plate 5.17 Contour furrows in Kitui Province, Kenya

Plate 5.18
Animal-drawn ridger to form contour ridges (H. Lal)

Plate 5.19
A tractor - drawn ridger - blade (H. Lal)

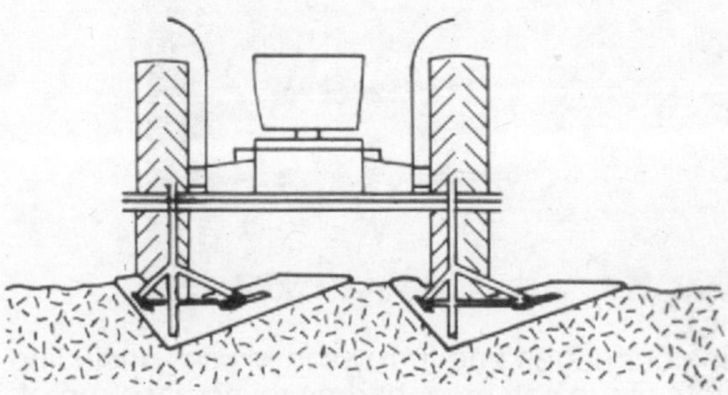

Figure 5.7
A non-Symmetrical furrow former developed in Mexico (after Garduno 1981)

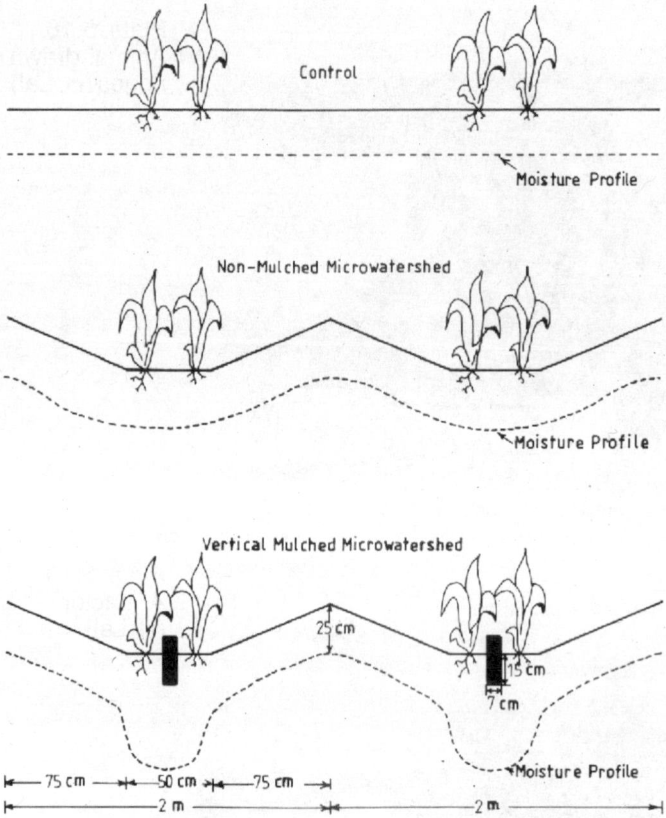

Figure 5.8 Studies of the effect on soil moisture of contour ridges and furrows, also the effect of mulching in trenches in the furrow (Fairbourn 1975; Gardner 1975

Plate 5.20 Field- trials of contour ridges and furrows in Mexico, including various treatments of the catchment areas to increase run-off (S. Gavande)

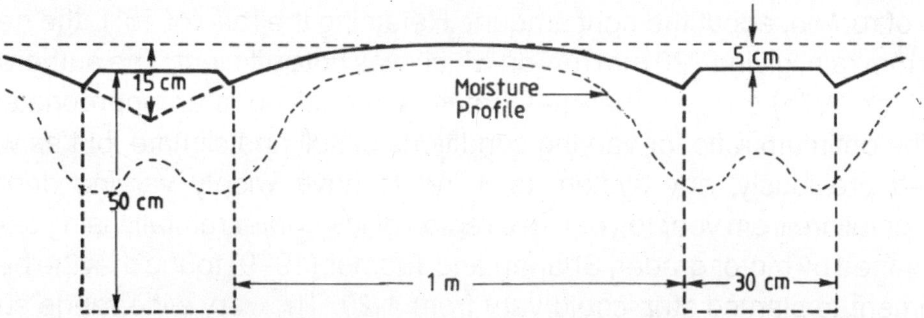

Figure 5.9 A proposed type of ridging for Botswana where a hard crust forms on the undisturbed ridges, giving higher run-off onto the cultivated strips

Treatment of the run-off area with additives, as in Plate 5.20, is unlikely to be economic, but it may be possible to take advantage of naturally occurring soil properties to increase the run-off. A case in point is Botswana where the sandy soils known as the 'hardveld' become hard and crusted after heavy rain. A method proposed there by Nilsson consists of forming permanent low ridges about a metre wide which are left to develop this hard crust, while ploughing, cultivating, and planting are restricted to narrow strips of 30 cm (Figure 5.9). This system has the advantage that after it has been established, no further maintenance is required.

The concept of strip tillage on the contour with ripping to encourage infiltration was tested in Tanzania by Macartney et al. (1971), and in Botswana, Willcocks (1981) showed that precision strip tillage was more effective than other cultivation methods, particularly if the strips could be consistently aligned along the previous crop rows.

The practice of cultivating to leave ridges and furrows on the contour without any additional land shaping was discussed in Section 4.2. It appears to be more common for maize and sorghum to be planted on the ridges, and small grains in the furrows. The practice of drilling wheat, barley, and rye into small furrows is reported from Poland by Wollen (1974) and is common in the wheat growing areas of Kazakhstan, USSR.

In low rainfall areas a large ratio of catchment to cropped area is required, but it is not easy to design a system which will give the best result for all variations of annual rainfall. In some experiments in North America the ratio of 33:1 was tried, and with 190 mm of rain gave 530 mm of run-on to the farmed area, which was more than required and more than could be absorbed. The next year the ratio was reduced to 15:1, but the rain was greater at 246 mm and with a higher run-off coefficient gave

390 mm of run-on, about the right amount. Retaining the ratio of 15:1, the next year had 140 mm rain, giving 220 mm run-on which was not sufficient. The authors (Morin and Matlock 1975) suggest that mathematical modelling is an appropriate tool to design the optimum ratio for varying conditions of soil and climate, but as we have discussed previously, any system is going to have widely varying degrees of success or failure from year to year. In a region of low winter rainfall, using saw-tooth ridges shaped by motor grader, Shanan and Tadmor (1979) found that the best ratio of catchment to planted strip could vary from 4-20. The very wide range suggests that trial and error may be as effective as mathematical modelling.

5.3 WATER SPREADING (THE USE OF RUN-ON AREAS)

Here again we have a wide variety of methods for different conditions. The soil type influences the method in several ways. Clay soils have low infiltration rates and high moisture storage capacity, so they are suitable for deep flooding with subsequent cropping, e.g. the tanks in India described in Section 5.3.3. Sandy soils have quicker infiltration and lower storage which tends to go with the diversion schemes of Section 5.3.2. A deep soil can absorb larger amounts of water, while a shallow soil may need the provision of overflow outlets to avoid drowning the crop.

The climate too; affects the method. Mediterranean climates have winter rainfall and low evaporation which may make possible run-off farming with mean annual rainfall (MAR) down to 100 mm as in the Negev desert in Israel (Evanari et al. 1982) and in Tunisia (Damagnez 1979). But with tropical summer rainfall and high evaporation, water harvesting is most likely to be useful in areas with more than 250 mm.

The probability of high-intensity rain influences the method. In the Negev most rain falls in light showers so the objective is to maximize the run-off (discussed in Section 6.2). In tropical Africa the problem is to handle the sudden floods from violent summer thunderstorms.

If the rain and the growing period do not coincide, the object will be to store water for later use. If the rainfall and growing period are the same, the desired result will be several applications of water during the period of crop growth.

5.3.1 Natural Run-off

There are many examples of traditional use of naturally occurring run-off to augment rainfall in areas where the rain alone is not sufficient for growing crops. Some authors have described this as "exploiting micro-niches in the environment".

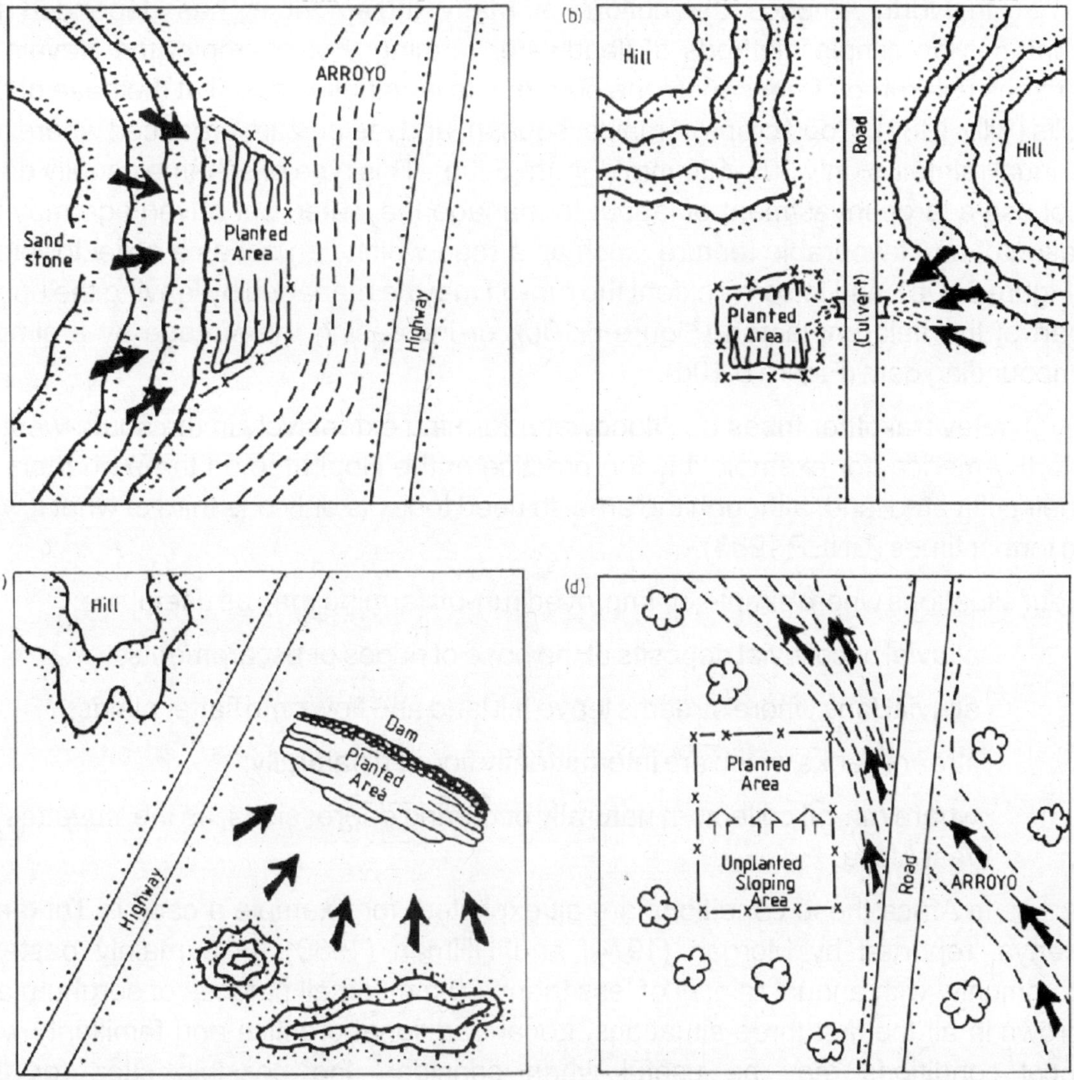

Figure 5.10 Floodwater farming in Arizona by Navajo Indians (Billy 1981)

(a) Farming with natural floodwater from a sandstone ridge
(b) Floodwater collected by a road drain and channelled through a culvert
(c) Planting after floodwater has been retained by a low dam
(d) Planting the lower half of a field which receives floodwater from the upper part

In North America the culture of many Indian tribes has depended for centuries on simple methods of floodwater farming. For example, the Navajo in Arizona use run-off from sandstone outcrops to water alluvial soil at the base of the hills (Billy 1981). Good crops of maize, squash, and melons are produced where the annual rainfall is only 300-400 mm (Figure 5.l0a). This type of scheme usually does not use a large investment of labour to manage the water, but advantage may be taken of any favourable feature, such as a road which can act as a collector drain (Figure 5.lOb), and to some extent the run-off may be managed by leaving the upper part of the field unplanted (Figure 5.10c), or increasing the storage by a simple impounding dam (Figure 5.10d).

Several other tribes use floodwater farming extensively in the south-west of North America, for example it is the practice of the Hopi tribe on three quarters of their cultivated land, although the area so used today is only one third of what it was in former times (UNEP 1983).

Four situations where simple, unimproved run-off farming may be useful are:

· alluvial or colluvial deposits at the base of ridges or escarpments;

· alluvial fans where streams leave hill land and flow onto flatter plains;

· stream banks which are intermittently flooded naturally;

· where run-off collects in naturally occurring depressions, as the 'cuvettes' in West Africa.

In Africa these conditions are all exploited, for example a case in Turkana, Kenya, reported by Morgan (1974) and Hillman (1980). In a mainly pastoral community with annual rainfall of less than 200 mm, small patches of sorghum are grown in all the first three situations. Local traditional practice and familiarity with local conditions may be useful when consider- ing possible sites for the development of more ambitious run-off farming schemes, and this was the case in the Turkana example (Cullis 1985).

Where the rainfall is sufficient for general cropping (and this usually means more than about 500 mm), it may still be possible to make use of run-on areas by increasing the cropping intensity. In the use of con- servation bench terraces (Section 5.2.3), separate rotations are used for the run-off area and the run-on terrace. The same principle is applied to natural run-off in Morocco (Kutsch 1982 reported in Pacey 1986). When growing wheat or barley, a plant density of 70-90 plants/m^2 is used on land receiving rain only, but the plant density is increased to 90-

120 plants/m^2 in depressions which collect run-on. This does entail an element of chance, described by Pacey as "the deliberate reduction of plant density results in some sacrifice of yield in seasons of ample rainfall, but it makes for greater certainty that there will be useful grain production in a drought year."

Two examples are reported by Kovda (1961) of the use of naturally occurring run-off in arid areas of the USSR. 'Kair' farming is the name given to cropping on flood terraces of the large rivers of central Asia where the soil moisture is partly the result of surface flooding, but also by lateral seepage from the river. Another example is 'khaki' farming in Turkmenistan where run-off from rain on higher mountain slopes inundates gentle slopes in the plains. The flooded fields are left to dry out and then cultivated and planted as soon as is practicable. A similar practice is reported from Ethiopia by Carr (1979). In the Woito and lower Omo valleys, as soon as the annual flood recedes, the flood plains are cleared and planted to maize and sorghum, using heavy seeding rates which are thinned later.

5.3.2 Collected and Diverted Run-off

The previous Section describes some uses of natural or unimproved run-off, while this Section is about schemes where there is some element of manipulation or management of the land or the run-off. The difference is not important, because some systems use only very simple works. For example, the floodwater farming of the Hopi and Navajo Indians of northern Arizona, described in Section 5.3.1, is very similar to that of the Papago Indians of southern Arizona who make simple barriers of wooden posts and woven brush to spread natural floodwaters more evenly (Nabhan 1979).

It is not important whether we consider run-off farming as a type of irrigation. In the case of terraced wadis and other water management schemes in valley bottoms, man sets out to manage both the land and the water. Farming valley bottoms is a very ancient practice and has been well documented in Tunisia and in the Negev desert in Israel. In the Matmatas of Southern Tunisia the system is called 'Gessours'. In an area with only 100-150 mm of rain, rows of olive trees are planted across the width of the valley bottom near a minor barrage, which slows down the floods, and result in deposition of silt with a terracing effect (Plate 5.21). Cereal crops are grown on each terrace upstream of the olive trees. This is similar to the terraced wadis in the southern Negev in Israel (see Plate 5.22). In some cases these ancient schemes are partly cultivated today by the mainly pastoral Bedouin inhabitants (Plate 5.23).

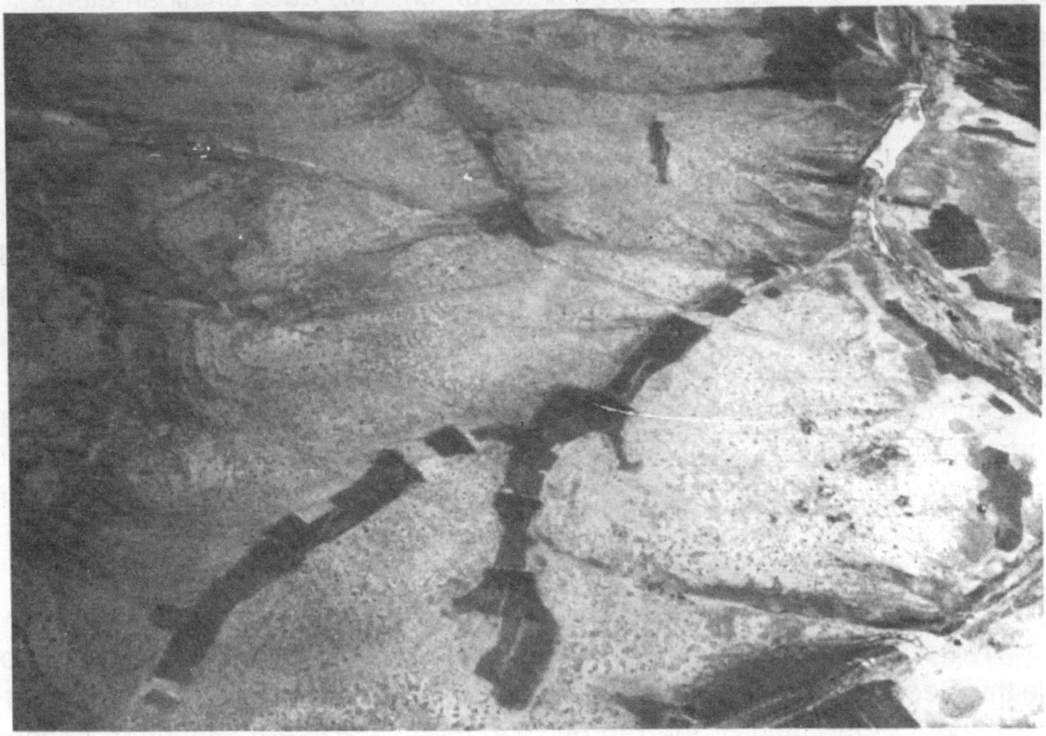

Plate 5.21 'Gessours' on terraced wadis in Tunisia (J. Damagnez)

Another example of carefully controlled water use in valley bottoms comes from Colorado (Mickelson et al. 1965). Level pans of one to three hectares are formed in broad valleys where slopes are less than 3 percent. The rainfall is 400 mm, sufficient for some rather unreliable cropping of grain or forage sorghum. The cropped land yields 5 to l0 percent of the rainfall as run-off from the heavy summer storms. The run-off from 150 ha of cropland is spread onto the first level pan, and when the depth reaches more than 100 mm the surplus spills over to the second pan and so on. During the season an additional 200 mm of water results in good crops on the pans, so that the installation costs can be repaid in three to five years. In most schemes involving flooding of crops, it is usual to provide outlets for emergency overflow, and possibly for draining as well, because growing crops appear to be less affected when there is a slow continuous passage of water than if the water is stationary. Also most crops only tolerate complete saturation for a short time.

Plate 5.22 Ancient terraced wadis in Southern Negev, Israel, now abandoned (Mrs. L. Evenari)

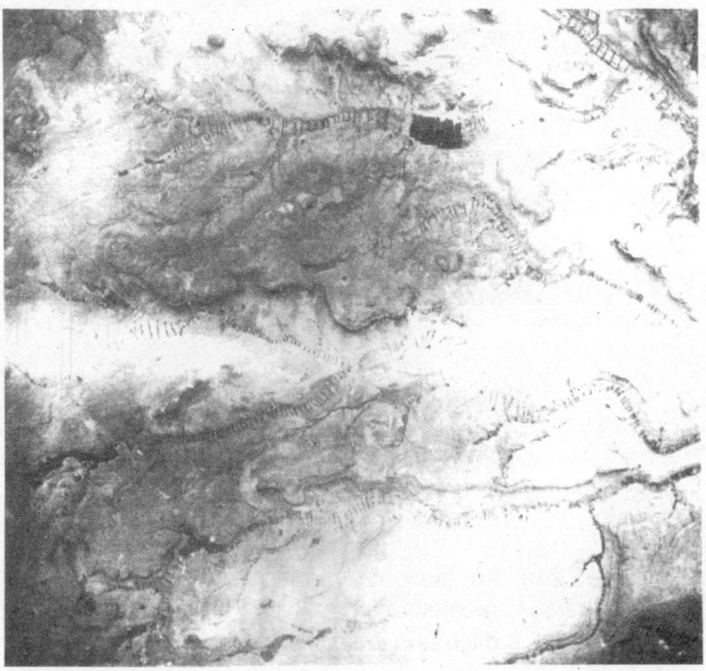

Figure 5.23
Widespread ancient terraced wadis in the Negev, Israel; most are abandoned but the dark terraces are cultivated by Bedouins (Mrs. L. Evenari)

Some of the ancient Negev valley floor cultivation systems used only spate run-off down the valley, but most of them depended also on the collection of run-off from the surrounding hills. Some examples are shown in Plates 5.24a to d. Some had sophisticated methods for inducing run-off, collecting it in channels and distributing it to the various terraced fields. A fascinating programme of research has been conducted for twenty years by the universities and research institutes of Israel and has been splendidly recorded (Evenari et al. 1982). The surface management to induce run-off included collecting and piling all the loose surface stones into mounds, and shaping the surface into ridges which are very similar to the roaded catchments used nowadays in Australia (see Plate 5.25). The hydrological effects of removing the stones are explained in Section 6.2.1. and the effect of building ridges in Section 6.2.2.

Plate 5.24a An ancient run-off farm in the Negev desert has been reconstructed at Avdat, Israel, using the old run-off collection ditches in the background (Mrs. L. Evenari)

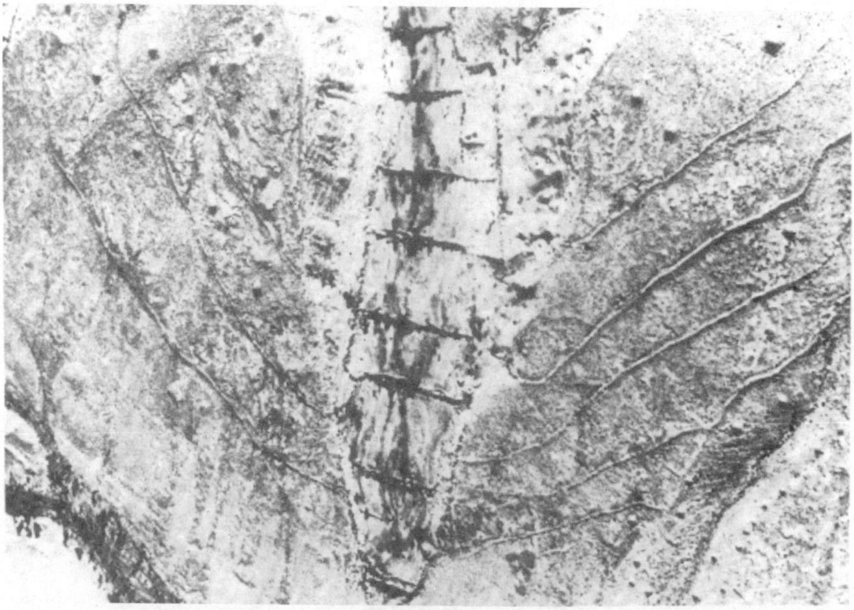

Plate 5.24b A farm unit near Shiuta, Negev, Israel, with conduits leading run-off to the cultivated terraces (Mr.s L. Evenari)

Plate 5.24c Many separate run-off farming units, each with its own run-off collecting channels, Israel (Mrs. L. Evenari)

Plate 5.24d A large run-off farm unit, with tree-grown homestead area on the right (Mrs. L. Evenari)

Plate 5.25 Shaped gravel strips increase run-off and lead it to the cultivated fields in the valley, Negev desert, Israel (Mrs. Evenari)

The collection systems in the Negev show that the hydrologic principles were fully understood. Stone-lined conduits were used to take the water quickly down to the farms with minimum loss by infiltration and evaporation, and the catchments for these drains were long and thin so that rain was collected even from the low intensity winter rainfall common in this area. The distribution systems were also simple in construction but sophisticated in concept, and allowed the water to be transferred and diverted around the farm quickly and efficiently with minimum labour requirements.

As a result of the successful studies in the Negev desert, trials of run-off farming were carried out in the Khost plain, in Paktia province of Afghanistan. The terrain is similar to the Negev; stony barren hills where run-off is collected in channels and led down to the plain. Level fields, surrounded by an earth bank, are laid out in shallow steps so the water is passed down-slope from field to field (Plate 5.26). After the soil profile has been saturated a cereal crop is grown on a combination of subsequent irrigations if there is rain, and the stored moisture in dry periods.

Another ancient collecting system called 'meskats' comes from the Sousse region of Tunisia with a rainfall of about 300 mm. Olives are grown on the better soils at the bottom of slopes. On the upper slopes, the soils have a tendancy to surface crusting and yield run-off into long thin catch- ment areas feeding the water down in channels to the olive groves (Plate 5.27).

These methods, developed for regions with winter rainfall, may not be suitable for the semi-arid tropics. Summer rainfall is more intense, so in general the catchment area required will be smaller, and provision for the safe passage of exceptionally high rates of run-off must be included. Figure 5.11 shows a system which has been tried on a small scale in two areas in Kenya, in Baringo by the Baringo Pilot Semi-Arid Area Project (BPSAAP) and in Katui by the Arid and Semi-Arid Lands Project (ASAL). Although not yet adopted to any great extent by the local farmers, the national policy of land allocation to what has traditionally been a mobile pastoral community, will probably lead to a stronger motivation for regular cultivation on permanent plots, which will not be possible without some form of water harvesting. Ratios of catchment area to cultivated land of up to 5:1 are expected to be suitable. For the system to work well, the cultivated area should be levelled so that there is little or no lateral slope along each terrace, and only a gentle slope across the terrace to avoid erosive flows. Stone-packed spillways built into the earth bund allow surplus water to pass progressively downslope to lower terraces. The danger of damage from excessive floods should be avoided, not by trying to control the water once it has reached the cultivated area, but through the provision of overflow by-passes set into the bank of the collection drain. There is no doubt that such a system can be made to work on an experimental or demonstration basis, and the question is whether the pressures of social change will make it sufficiently attractive to people unaccustomed to a cropping system which requires this degree of precision and maintenance.

Another form of water spreading in USSR known as 'liman' irrigation is described by Kovda (1961). This consists of what are described as contour ramparts across the slope, perhaps an apt description since they are large structures up to two metres high and four or five metres wide (similar to the 'murundum' water conservation structures used in Parana State in Brazil). Sometimes drain holes are used to pass the water through to lower levels, although an open spillway would appear to offer a simpler and more effective way of controlling the water movement.

The diversion of run-off water onto prepared level terraces is another ancient and widely used method. This is the basis of growing irrigated paddy in humid areas where water supplies are assured, and the same principles may be used to obtain benefit from intermittent water supplies in semi-arid regions. For example, in the mountainous dry areas of northern Pakistan, in the district of Gilgit and the Hunza valley, agriculture is entirely dependent upon the diversion of springs, streams, and melt waters from snow and glaciers, onto terraced alluvial outwash fans which are accurately called oases, Plate 5.28. Some amazing construction feats have been used to carry the 'khuls' across unstable scree slopes, and laboriously cut by hand out of rock faces, Plate 5.29.

Another interesting example of diversion onto terraces comes from the Manakhah region of the Yemen Arab Republic (Damagnez 1979, Vogel 1985). In an area of annual rainfall of 400-600 mm, level bench terraces are constructed on steep slopes up the hillsides, with rock risers 2-3 m high and the terraced fields 2-l0 m wide (Plates 5.30 and 31). Run-off from the uncultivated and unterraced upland is carried to the terraces in diagonal ditches reminiscent of those in the Negev. Stone-lined canals carry the water progressively down through the terrace system, sometimes with the canals running underneath the terraces. This sophisticated and efficient system is presently falling into decay (see Plate 4.18) because of labour migration to the oil producing countries of the Middle East. A similar system in the Yemen Arab Republic, called 'sayl' irrigation, is the diversion of flood water out of wadis onto levelled terraces.

Plate 5.26 Run-off farming on terraced fields in Afghanistan using run-off from the stony barren hill in the foreground (D. Gebauer)

Plate 5.27 In the Sousse region of Tunisia, collecting drains ('meskats') lead run-off down to olive groves (J. Damagnez)

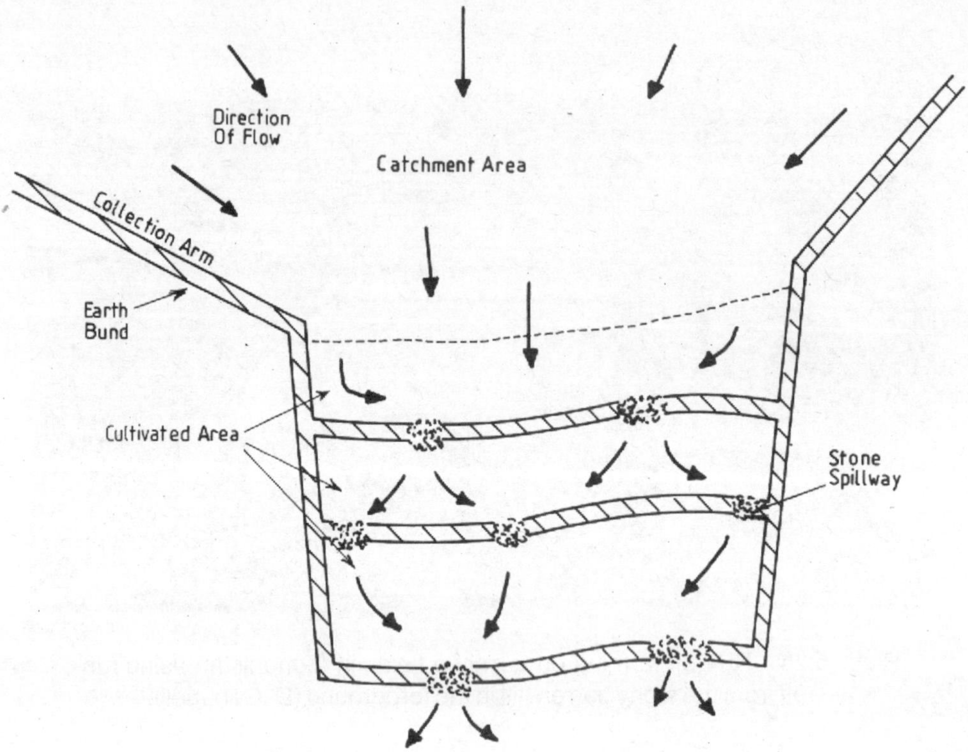

Figure 5.11 A run-off farming system from Kenya (Charnock 1985

The management of grazing land is discussed in Chapter 7, but a water spreading system from Australia is worth mentioning. In New South Wales, Australia, run-off occurs from eroded and degraded ridges and is absorbed on the flat land below the ridge, but produces only inedible scrub vegetation, sometimes called 'mulga' after the dominant shrub, Acacia aneura. Two approaches are used to make more efficient use of the rain- fall. One is to encourage useful grass growth by retaining it on the ridges using short contour furrows as described in Section 7, (Cunningham 1967 and 1975). The other approach is to catch the run-off in drains and lead it to where it can be spread on land for fodder crops and pastures. The collector drain has the bank on the lower side; the spreader drain has the spoil bank on the upper side so that the water spills evenly over the downhill edge of the channel (Figure 5.12). During large run-off flows, water also collects on the upstream side of the spreader channel, so gaps are provided to allow this water to pass through into the spreader channel (Quilty 1972).

Plate 5.28 An oasis in the Hunza valley, northern Pakistan. The 'khuls' appear as horizontal lines on the hillsides

Plate 5.29 The supply canals or 'khuls' across steep slopes

Plate 5.30 Terraced cultivation using hill run-off in the Yemen Arab Republic (J. Damagnez)

5.3.3 Inundation Methods

Some examples of naturally occurring short-term flooding were described in Section 5.3.1, mainly on flood plains and in valley bottoms. This Section describes systems where flood waters are impounded and retained long enough to saturate the soil so that a crop can be grown on moisture stored in the soil.

Plate 5.31 Run-off farming in the Yemen Arab Republic (J. Damagnez)

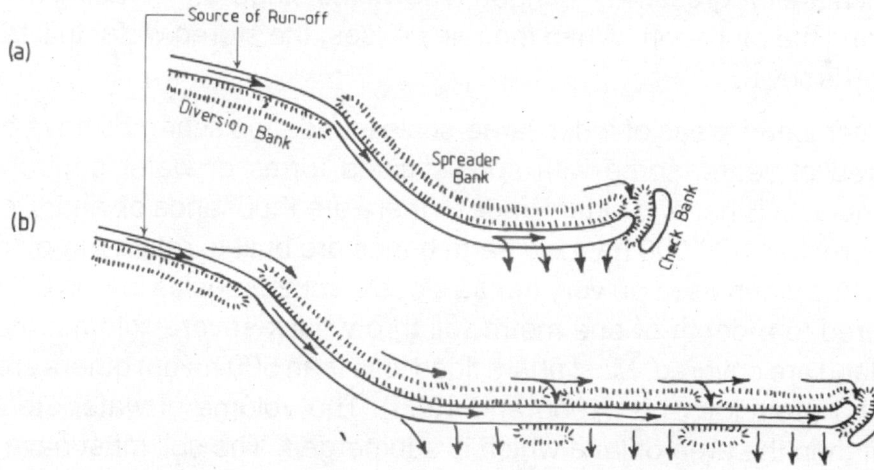

Figure 5.12 Run-off diversion and spreading in Australia (from Quilty 1972)
(a) for small flows, the spreader bank is continuous
(b) for larger flows, run-off will build up above the spreader bank but is passed through gaps to the channel. In both cases, there is a check bank at the end of the diversion to check the flow of diverted run-off

Simple systems on a small scale are used in the Sudan. On gently sloping land embankments known as 'teras' collect and hold surface run-off, and after the water has soaked into the soil a short duration millet is planted and matures in eighty days (Pacey 1986 p 131). In the semi-arid north-east of Brazil, the government research organization CPATSA in Petrolina has studied a similar method, and produced designs for the optimum shape and size of the bunds. There are also recommended cropping methods to make maximum use of the stored moisture, with variations in the cropping pattern depending upon the amount of rainfall and the amount of run-off stored (EMBRAPA 1981). The inundated areas are known as 'vazantes'.

In India the use of contour bunds to retain run-off has a mixed story of success and failure. When used mainly as a soil and water conservation measure in areas of reasonable rainfall (750 mm - 1250 mm) on medium-deep vertisols, it was found that the disadvantage of waterlogging in the vicinity of the bund both uphill and downhill exceeded the advantage of increased cropping from the stored moisture in a dry season (ICRISAT 1975). But in areas of lower rainfall the method has been more successful, for example in the Siwana district of western Rajasthan. With a rainfall of 250 mm or more, low contour bunds only 0.3 m high divide the land into strips which are progressively inundated down the slope with flooding to a depth of 0.2 m during the monsoon. When the rain ceases, the stored water infiltrates and a winter crop is sown.

In semi-arid areas of India, large-scale inundation schemes have been used for hundred of years, some with sophisticated forms of water control and land management. In Bihar and Uttar Pradesh there are thousands of 'ahars' covering a total of more than 800 000 ha. Low earth bunds are built to retain run-off during the monsoon and when used on very gentle slopes, sometimes as low as 0.01 percent, water stored to a depth of one metre will throw back several kilometres, so large areas of land are covered. Most ahars flood less than 500 ha but others are as much as 4000 ha (Kolarkar et al. 1980 and 1983). The volume of water stored is less important than the area of land which is submerged. The soil must have sufficient depth and moisture-holding capacity to store enough moisture to carry a five month crop which is grown in 'rabi', the dry winter season, after the end of the summer monsoon season, 'kharif'.

The retaining bank is usually not more than 3 m high and may extend several kilometres on the contour, with some examples up to 10 km. There is usually a wide waste-weir to serve as emergency spillway, usually stone pitched with a crest level one metre below the top of the bund to give one metre freeboard for wave action. The emergency spillway is required partly because of the danger of a heavy storm occurring when the tank is nearly full, and also because the ahars are usually built in a series down the slope and the failure of any one would imperil those lower down the slope.

It is also usual to provide metal sluice-gates set in concrete structures to allow for quick emptying when necessary at the end of the rains. The water released may be collected in the next ahar down the slope. Sometimes pipes are laid through the bund if some of the discharged water is to be used for irrigation.

The main crop is winter wheat planted as soon as the flooded land has dried out sufficiently, and then grown out on the stored moisture. Sometimes a subsidiary crop is taken during the summer monsoon while the land is flooded, using a variety of rice known as 'floating rice' because it can grow up through the standing water. This technique of taking one crop of floating rice followed by a second crop after the water has receded is also practised in Thailand.

A secondary advantage of this system is that the soils in these semi-arid areas often have a tendency to salinity, and this is controlled by the regular leaching by the inundation water. The infiltration of considerable amounts of water may also have the effect of raising the general level of the water table near to the ahars, with improvement of the supply in shallow wells.

Another form of inunadation farming is used in the Jaisalmer district of Rajasthan (Kolarkar et al. 1983). Here the topography is more broken, and the method is to build an earth bund across the valley plain to catch and store run-off and silt from the surrounding barren hills. These are called submergence tanks or 'khadins' (Figure 5.13). In the district of Jaisalmer there are more than 500 such tanks with an inundated area of more than 12 000 ha. A catchment ratio of at least 15:1 is usual and the tanks are designed to fill from rainfall of 75-100 mm, in an area where the annual rainfall is 165 mm with great variability. The main construction features are the same as the ahars, that is a wide emergency spillway and sluices to release the water. The depth of water stored varies from 0.5 to 1.25 m and it usually disappears from seepage and evaporation by early November, when the winter crop is sown, either winter wheat or the grain legume Cicer arietinum, (Bengal gram or chickpea). Similar systems of inundatiion farming Pakistan are known as 'sailabas' and 'kuskabas'.

Secondary benefits of most inundation systems are the leaching of salinity by the stored water and the improvement to the yield of wells downstream.

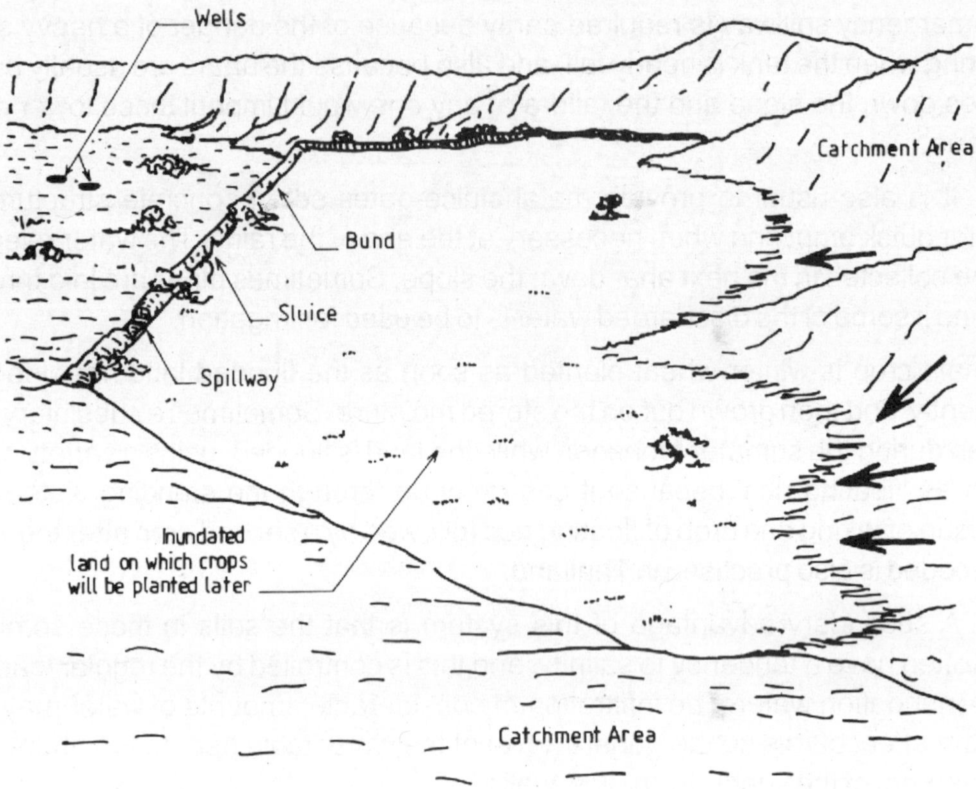

Figure 5.13 Sketch of a 'Khadin' or submergence tank in India (from Pacey 1986)

5.3.4 Flood Diversion

The diversion of run-off water was discussed in Section 5.3.2. This Section is concerned with diversion and spreading of floods and spate flows.

Diversion of flood water from its channel usually involves some form of structure, a barrage or weir to divert the water. For small schemes, simple diversions may be constructed each year using stones and boulders, perhaps in wire netting, or poles and brushwood where these are available. It may be sensible to expend only limited effort on building such struc- tures where it is probable that they will be periodically destroyed by flash floods, and an example from Yemen was quoted in Section 4.1.4. The floods are also likely to damage the conveyance channel unless it is pro- vided with safety devices to spill water picked up in large floods. To avoid these problems a method has been widely used in India for centuries, the inundation canals, or 'pynes'. A canal is excavated in the river bank so that the level of the canal bed is considerably higher than the bed level of the river. During low flows the canal is dry, but when the flood level rises to that of the diversion canal it starts to flow. There is no need for any structure in the stream or river channel. This system is widely used in the Sind province of Pakistan to collect water from the Indus river and its tributaries during the flood period from April or May to September. The grade of the distribution canals is slightly flatter than that of the river bed to increase the command of the diverted flood water. This also means that there is heavy deposition of sediment parti- cularly at the beginning of the canal, and regular clearing is essential. Some control is also required of the diverted flood water and this is usually done by a regulator several kilometres from the river. Subsequent control and distribution of the water is the same as for any surface irrigation scheme, except that the flow is ephemeral and used for heavy inundation so less precise control is required.

The other possibility is to divert the flood water out of its channel by raising the water level through some form of weir or barrage. These may be temporary and expendable in simple schemes, or permanent structures. The problems to be overcome by all flood diversion schemes are:

- damage to the diversion works during flash floods;
- control and spreading of the diverted flood water;
- deposition of sediment carried within the flood water.

In some cases the last one may be considered beneficial, as in the flood irrigation in the valley of the Nile, and flood irrigation in China from silt-laden rivers, where this is known as 'warping' and considered to be part of the process of maintaining fertility. The mica-laden melt waters from the glaciers of the Himalaya are used to build up the physical volume of soil by spreading the water onto gravel beds. On the other hand, too much deposition of sediment can lead to undesirable change of soil texture, or to reduced yield of grassland.

Another example of large-scale diversion of flood waters comes from the Yemen Arab Republic where the valleys discharge into the Tihama desert which bounds the Red Sea, using elaborate diversion structures as shown in Plate 5.32 (Damagnez 1979). Another variation of large-scale use of flood water is illustrated in Plate 5.33, where structures are used to control the flood water after it has spilt naturally from the River Niger. Plate 5.34 shows an example of using spate flood water to inundate an olive orchard in Sfax, Tunisia, in an area with an annual rainfall of 150 mm.

Plate 5.32 Flood diversion structure in the Tihama Plain, Yemen Arab Republic (J. Damagnez)

Plate 5.33 Floodwater control structure, plain of river Niger, Mali (J. Hartley, Oxfam, EuroAction)

Plate 5.34 Inundation of an olive orchard in Sfax, Tunisia, using diverted flood water (J. Damagnez)

The diversion of flood water can also be practised on a smaller scale as illustrated in Figure 5.14 showing an interesting project in Tigre, Ethiopia. Water is first diverted from the river and then distributed using a simple system of structures constructed from local stone, shown in Plates 5.35 and 5.36. This is a development project sponsored by the UK charity War on Want, and is continuing. A similar scheme on a pilot project in Turkana, Kenya, is shown in Figure 5.15. This pilot project was successful for a short time but was soon abandoned, first because of siltation at the diversion weir and later because the river changed its course. A basic problem with the introduction of schemes like this is that permanent cultivation on settled sites is not part of the traditional farming system, so there is little interest, and little motivation for the regular maintenance which is required. However there are some indigenous water spreading schemes, and examples from the Sahelian belt are reported by Pacey (1986), stretching from Southern Mauritania and Niger to Sudan and Somalia. Other examples of indigenous soil and water conservation methods in Africa south of the Sahara are given in a study commissioned by IFAD (1986).

Examples of the diversion of flood water to improve grazing are discussed in Chapter 7.

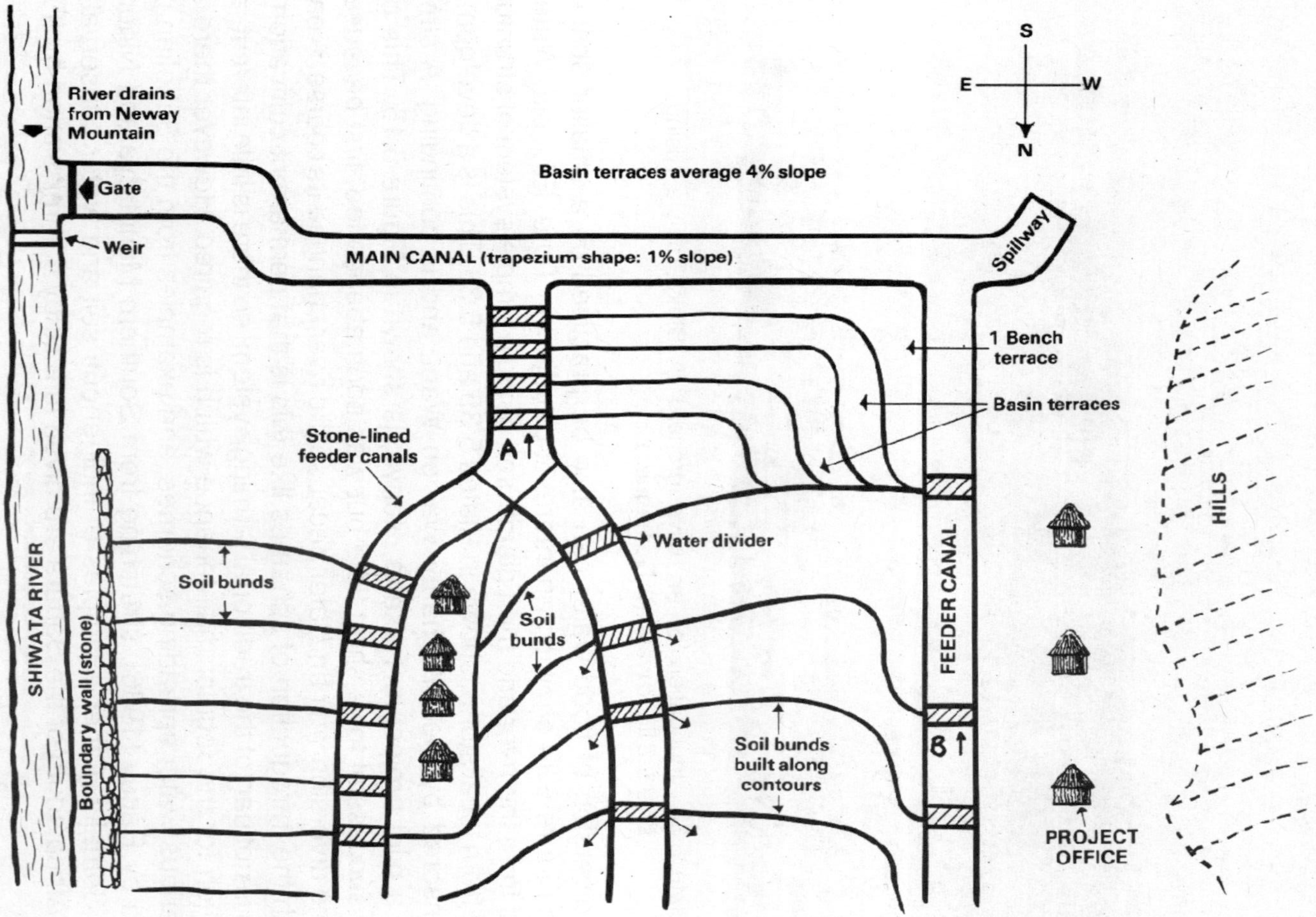

Figure 5.14 A small flood diversion schem in Tigre, Ethiopia (Bennett 1985)

Plate 5.35 Flood diversion structures, Tigre, Ethiopia, taken at point A on Figure 5.14 (J. Bennett, Andes press Agency)

Plate 5.36 The feeder canal taking water to the lower terraces, taken at point B on Figure 5.14 (J. Bennett, Andes Press Agency)

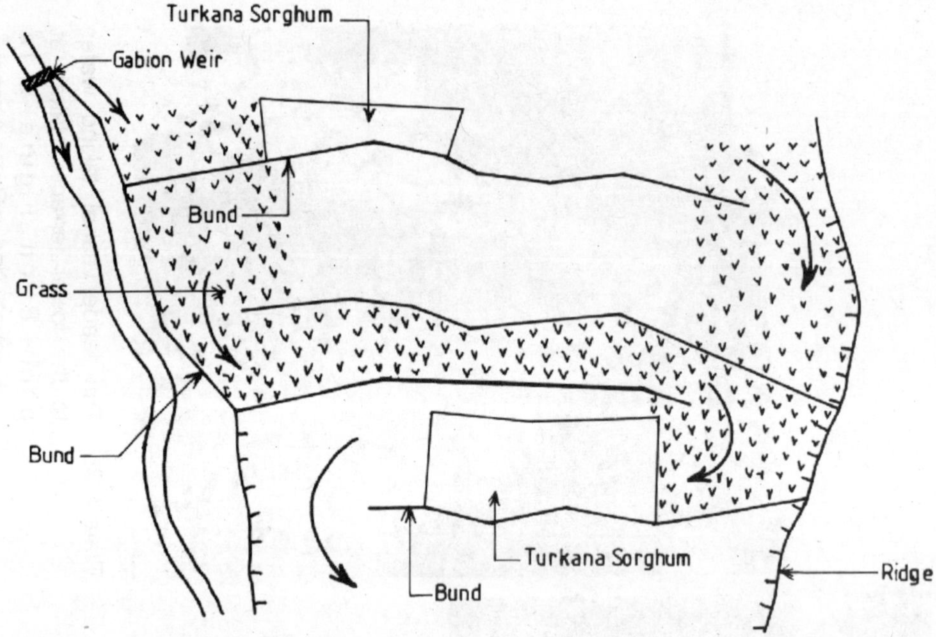

Figure 5.15 A pilot project in Turkana, Kenya, Diverting flood water onto pasture and crop (Pacey 1986)

5.4 SURFACE DRAINAGE

The main problem in semi-arid climates is shortage of water, but there are still occasions when at particular times or on special soil conditions the problem is too much water, and some form of drainage is required.

A soil where some form of water control can be important is the vertisols, heavy clay soils, usually dark, also called cracking black clay soils, cotton soils, and self-mulching clays. They are also known collo- quially as 'one-day soils' because they are difficult to cultivate when dry and hard, and also unworkable when saturated and sticky, but they do have a short window while drying out when the moisture is just right for cultiva- tion. The physical properties causing this are a high clay content, with the clay mineral being montmorillonite or illite, which, because of their 2:1 lattice structure, expand when wet, and shrink and crack when dry. A characteristic surface feature is the development of an uneven surface with raised mounds and shallow depressions caused by repeated cycles of wetting and drying and known in Australia as 'gilgai'.

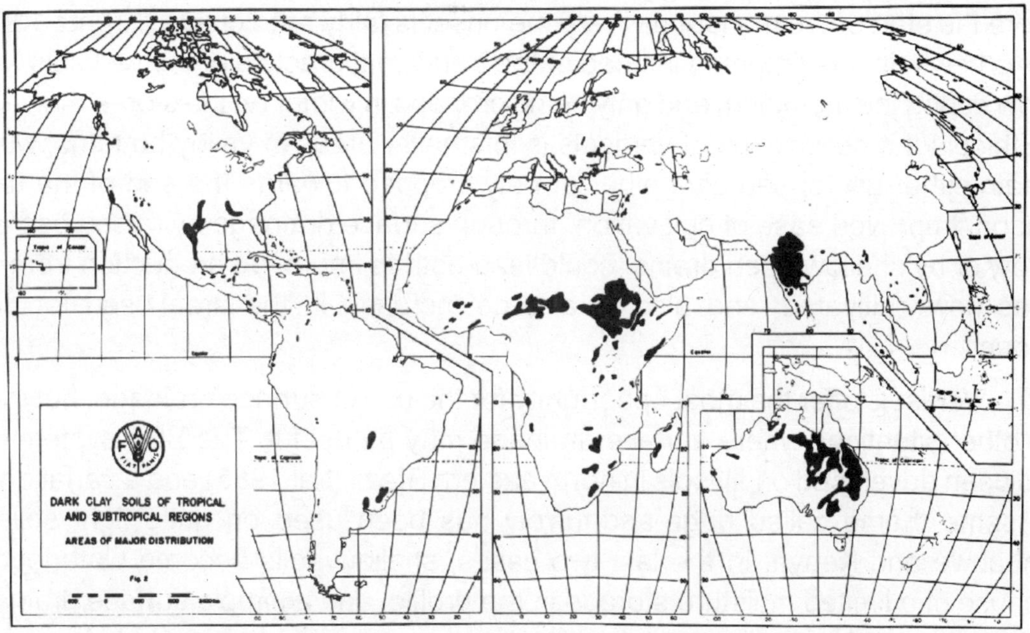

Figure 5.16 The main areas of dark heavy clay soils in the tropics and subtropics (Dudal 1965)

The occurrence of vertisols is mapped in Figure 5.16 from Dudal (1965). The two main kinds of formation are the large uniform areas derived in situ from extensive volcanic flows of basalt, or smaller local- ized alluvial deposits in valley bottoms as in Ethiopia, Zimbabwe, and Zambia. In both situations, vertisols have considerable economic signifi- cance because they are a land resource with under-utilized potential owing to the difficulty of handling them.

One technical solution is the installation of subsurface drainage, and this was shown to be effective more than twenty years ago in Zimbabwe (Robertson 1964). Pipes of concrete, fired clay, and bamboo were all effective but uneconomic because of the cost of materials and transport. Today, the most appropriate method would be the installation of plastic tube drains by the trenchless method. The techology is available, but a high return would be required to justify the installation costs. Mole drains would be cheaper but the extremes of wet and dry conditions would lead to a short life.

Surface drainage is the more practical alternative and has been thoroughly developed and tested at ICRISAT in India, using the broad bed and furrow method described in Section 5.2.1 (Krantz et al. 1978, 1981, and Swindale and Miranda

1984). In India, surface drainage can lead to the opportunity for early dry-sowing just before the start of the monsoon where the onset is fairly predictable (Krantz 1981). This a1lows double cropping instead of the traditional practice of the vertisols lying fallow during the monsoon, and only being cropped in winter on the stored moisture. In Ethiopia the occurrence of vertisols is in smaller areas in valley bottoms, which remain either uncropped altogether or only cropped towards the end of the rainy season. Improved ease of cultivation, through surface drainage by raised bed and furrow or by shallow open drains, could lead both to improved production from the areas now cultivated, and also to a large increase in the area used for crop production.

Vertisols offer the most opportunity for the use of surface drainage, but there are other situations where surface drainage may be useful. The BBF system has also been developed on alfisols in north eastern Brazil (Lal 1985) and a variation on the same theme called ridge-and-furrow has been used on quite light soils in Zimbabwe and Kenya. In the last two cases, shallow soils become waterlogged because of a limited moisture storage in the profile, and a simple surface drainage system can lead to better opportunities for cropping, or an extended season for grazing. An interesting but specialized need for surface drainage comes from Ethiopia in the growing of teff which is the main food grain

The seeds are minute and broadcast onto a very fine tilth, and can easily be washed away by any surface run-off. The traditional solution is small open drains on a very steep gradient to intercept any surface flow. The need to prevent surface run-off is considered by the farmers to override other considerations like soil erosion or conserving moisture, and there results the illogical situation, shown in Plate 5.37, where the traditional open drains have been cut through more recent systems of channel terraces.

Another example of surface drainage to solve a particular problem comes from Portugal where the situation is temporary waterlogging at the time of preparing or planting winter cereals. The method is called 'pnudivales' which are small channels constructed after seed-bed prepara- tion and before seeding. They are made with three passes of a tractor. First with a mouldboard plough to create a furrow, followed by two passes with a mounted ridger to produce a shallow open drain between 3 and 4 m wide and 0.25 m deep at the centre. This allows subsequent drilling, cultivating, and harvesting in straight lines ignoring the open channels. These are set out on a gentle gradient of 0.6 to 0.8 percent (Sims 1983).

Plate 5.37 Steep diagonal surface drains of traditional style have been cut through modern graded channel terraces to increase surface drainage

5.5 OTHER SOURCES OF WATER: SNOW, DEW AND MIST

5.5.1 Snow Management

In high latitudes, some or all of the precipitation may fall as snow, particularly in the northern hemisphere, and in semi-arid regions there may be the possibility of increasing the effectiveness of the precipitation by management of the snowfalls.

Trapping snow during the winter to increase available moisture in the short summer is a regular practice on the semi-arid steppes of Eastern USSR in Kazakhstan. Some of the first snow is caught by the stubble of the preceding crop, or, in the case of summer fallow, by low lines of mustard planted for the purpose. As soon as 50 or 100 mm of snow has accumulated it is pushed into ridges by heavy-duty tractors with a V-shaped bulldozer on the front and an inward sloping half grader blade on either side, so that the snow is pushed into two windrows or snow ridges one on either side of the tractor. A typical design is 6 m between centres of the ridges with 3 m of untouched snow between. The ridges are up to 1 m high and 3 m wide, and serve as wind breaks to catch more snow between the ridges. In a year of very heavy snow the total snow pack between the ridges can be up to 600 mm but this is more than desirable since the melt water causes erosion problems; 400 mm is normal and more desirable.

The conservation bench terraces, described in Section 5.2.3, have the main purpose to hold on the level terraces the run-off from catchment areas on the uphill side. In the two most northern states of the USA where the method was tested, it was found that they also increased the amount of trapped snow, partly as a result of the terraces acting as a physical barrier, and also because of the stubble of the previous summer crop grown on the terraces (Haas and Willis 1968 on North Dakota, and Black 1968 on Montana). Two other methods of snow trapping have also been practised in Montana. Neff (1980) reports that contour furrows caught snow equivalent to an additional 22 mm of rain, and also reduced the run-off from winter rain and the snow melt in the spring. In eastern Montana, Saulmon (1973) used a standard snow fence (1.25 m high with vertical wooden slats) to trap snow on small catchments which provide surface run-off for small stock watering ponds. The additional depth of snow gave an increase of 100 mm of surface run-off.

5.5.2 Dew and Mist

The extensive literature on dew as a source of water contains many qualitative opinions about its importance or lack of it but few quantita- tive scientific studies. The most detailed study in semi-arid areas is that in the Negev by Evanari et al. (1982). They conclude that the annual total of dew may be significant in comparison with the annual rainfall. An example is a range of 25-35 mm from dew at Avdat where the mean rainfall is 150 mm. The dew formation is considerably more reliable than the rainfall, so that it is possible, as in 1962/63, for the annual dew formation at 28.4 mm to exceed the rainfall at 25.6 mm. This is very much greater than the amounts of dew formation precisely measured in England by Monteith (1957) who found that the average dewfall was from 2-5 mm. However, the desert dew is made up by a large number of very small amounts and not of any consequence as a water supply which can be collected, nor for any higher forms of desert plants, although it is used to some extent by the lower forms like lichens and xerophytic algae.

The Negev studies also showed that the piles of stones sometimes previously assumed to be dew collectors for the production of grape vines do not acquire enough dew on the surface of the stones for it to drip off, and that the mounds are in fact for the purpose of increasing surface run-off. There are also reports from China of a long history of growing melons in soil beds covered with a layer of gravel 100-150 mm thick and known as 'gravel fields for melon'. Condensation and dew formation dripping down to the soil is supposed to be the purpose of the gravel but its effect may perhaps be to maintain an even soil temperature or to reduce evaporation.

There are a number of examples of artificial collectors, such as plastic sheets being used to collect enough dew to keep alive forest seedlings in the desert in Israel (Gindel 1965), and another use of plastic sheets to produce emergency supplies of drinking water in Australia, but these do not have any agricultural significance.

There are also reports of aerial wells which supply drinking water in many dry areas, some with piped supplies, but the studies in the Negev would suggest that the process is more likely to have been the interception and condensation of fog or mist, or direct run-off, than dew formation (UNEP 1983). The story of dew formation and mist condensation is interest- ing, and full of contradictions, but the conclusion has to be that this is not a significant source for water for agriculture.

6. WATER HARVESTING AND USE

This Chapter describes methods where surface run-off is collected and stored in dams, tanks, or cisterns, for later use. Detailed informa- tion is given in the proceedings of two workshops, both in Arizona (Frasier 1975, Dutt et al. 1981). An excellent review of the state of the art is Hollick (1982).

6.1 HARVESTING METHODS

The subject may be considered in terms of:

- the purpose of the stored water, whether for domestic use, stock watering, supplementary or full irrigation;
- the type and quantity of storage; small amounts for domestic use or stock watering; below ground in cisterns or lined ponds or above ground in tank structures; larger quantities for irrigation, usually in ponds or reservoirs;
- whether the catchment is in its natural state or treated in some way to increase the run-off.

Stored systems nearly always modify the natural situation in some way, so there must therefore be some element of design. Factors to be considered in the design are:

- the amount of required water, which can be calculated from estimates of the numbers of people or stock and their daily requirements, or from the area and requirement of crop;
- rainfall, its frequency probability, and intensity, which combined with surface conditions will determine the run-off, and hencewhether some change is required to increase the run-off (discussed in Section 6.2);
- losses of the stored water through evaporation, seepage, or' leakage.

The number and range of the variables is so large that individual designs for every system would be laborious, but there cannot be standard solutions which will suit all conditions. The approach can therefore be either to use empirical experience, perhaps incorporated into design tables, as, has been done in Western Australia (Hollick 1975), or by using computer models, such as Frith et al. (1975) or Hanson (1975).

Degradation of the grazing lands in the Sahel has led to increasing concern about the necessity for maintaining the right balance between the available water supplies, available vegetation for browse and grazing, and the intensity of use by livestock. A number of examples are reported of undesirable changes to the ecosystem resulting from making more water available by the development of underground sources (in Botswana by M.L. and M.J. Odell, 1980, in the Sahel by Wade, 1974, and by Sinclair and Fryxell, 1985). This danger is less likely in the case of harvesting surface water, but it is still a possibility. One water harvesting method which can make use of all the available rainfall without risk of ecological upset is the buried membrane system developed by Shanan et al. (1981). The idea is to provide small cattle-watering points with run-off collected from catchments where a plastic membrane has been buried just below the soil surface. The amount of water collected is proportional to the rainfall and so is the nearby vegetation. If the amount of stored water is deliberately kept small it runs out before the vegetation, and nomadic herds would he moved onto the next watering point before the vegetation is damaged by overgrazing.

Two countries have a wide experience of water harvesting methods. These have been reviewed for North America, particularly Arizona, by Cluff (1975), and in Western Australia, by Laing (1981a).

6.2 TREATMENT OF CATCHMENTS

6.2.1 Natural Catchments and Minor Changes

Summer rainfall in the semi-arid tropics usually includes short- duration storms of high intensity which give high rates and amounts of run-off. This is partly because the high rainfall intensity can tempor- arily exceed the infiltration rate, and also because the infiltration may be reduced by the formation of surface crusts, caused by the breakdown of structure from the rain, and the high surface temperatures, as shown in Niger by Valentin (1985). In Burkina Faso, Roose and Piot (1984) measured annual run-off rates of between 30 and 40 percent on overgrazed and cultivated land, but in individual heavy showers the run-off was up to 70 percent. In Australia, even small falls of rain (less than 15 mm) produced run-off equivalent to over 50 percent of the rain from land with low density vegetation (Pressland and Lehane 1982). In semi-arid areas with winter rainfall there is a higher proportion of rain in small amounts and at low intensities. However, infrequent high intensity storms can lead to heavy damaging floods even with a winter rainfall of less than 100 mm.

The effect of the size of catchment area was discussed in Section 5.1.3. A small catchment will produce more run-off per unit area than a large catchment, and this is often made use of in the design and shaping of catchments, discussed in Section 6.2.2.

Significant increases in the yield of run-off can sometimes be achieved by simple treatment of the surface, for example the removal of vegetation or moving stones on the surface. In the northern Negev, Israel, with an annual rainfall of 250 mm, removing the vegetation decreased the infiltration rate by 40-50 percent and increased the run-off from 7 to 21 percent (Tadmor and Shanan 1969). They also report a similar effect from Utah where reducing the vegetation from 40 percent to 16 percent cover increased annual run-off by about 60 percent. The reason for this effect is partly that the plant cover gives protection against the rain-drop impact and lessens the crusting effect, while the plant roots and surface litter increase the infiltration. Another simple method which also has the effect of increasing run-off through less obvious, indeed unexpected, hydrological processes is the removal of stones from the soil surface. In the studies of run-off farming in the Negev desert, which have been previously referred to several times, one subject for research was the large number of mounds of stones or lines of stones which cover large areas of the catchments supplying water to the run-off farms in the valley. A superficial analysis of the hydrology would suggest that, if the soil surface is covered with a significant amount of stones, the run-off from this impermeable surface would increase the overall run-off. However, detailed studies with experimental run-off plots over a number of years gave two results which the authors themselves describe as 'astonishing'. These are firstly that with few exceptions run-off is less on steeper slopes, and secondly that clearing the stones from the surface and piling them into mounds increased the run-off. For an explanation of the complicated hydrological processes, interested readers are referred to Chapter IX of 'The Negev' by Evenari et al. 1982, and these have recently been confirmed in laboratory studies by Poesen (1986).

An obvious possibility for use as a catchment area is any naturally occurring rock outcrop, and there are worldwide examples of such use, particularly for domestic use and stock watering where the requirement is small quantities of good quality water. The geologically ancient Basement Complex formation often includes large granite outcrops which are traditionally used for domestic water supply in Zimbabwe (Richards 1972) and in Kenya (Plate 6.1). Similar schemes are found in Western Australia (Davies 1977, reported in Hollick 1982), in North America

(Chiarella and Beck 1975), and also in India and Sri Lanka. A simple collecting channel is required, usually a low wall of brick or masonry. The coefficient of run-off is not always as high as might be expected from bare rock surface, and can go down to less than 50 percent due to losses from evaporation, surface depression storage, and losses through fissures and cracks. Limestones are usually weathered and fissured, and may yield as little as 25 percent, while a steep solid granite may yield 80 or 90 percent. The yield can sometimes be increased by sealing the fissures with cement or bitumen.

Plate 6.1 Water harvesting from a granite outcrop, Kenya.
A stone wall about 20 cm high leads run-off to a storage dam cut of picture to the right

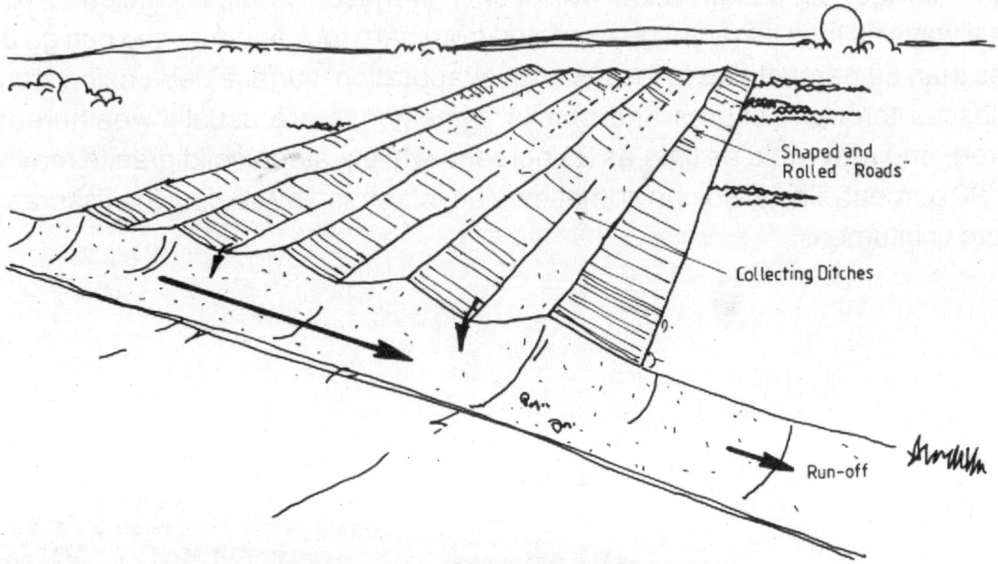

Figure 6.1 Roaded catchments are widely used in Western Australia for Harvesting run-off for domestic and stock - watering supplies

Plate 6.2 Roaded catchments, shaping the soil surface, Australia (I.A.F. Laing)

Plate 6.3 Smoothing and compacting the 'roads', Australia (I.A.F. Laing)

Plate 6.4 Erosion can be a problem on the 'roads' (I.A.F. Laing)

Plate 6.5 Two different layouts of roaded catchments, Western Australia (I.A.F. Laing)

Plate 6.6 Two examples of flat-batter dams, Western Australia (I.A.F. Laing)

6.2.2 Shaping Catchments

Two water harvesting methods widely used in Western Australia employ a careful reshaping of the soil surface, and usually some further treatment of the surface to improve run-off. The principle of 'roaded catchments' is shown in Figure 6.1 and Plates 6.2 to 6.5. Shaping the land into roads gives more run-off by decreasing the losses. Smoothing out the land reduces surface depression storage, and evaporation and infiltration losses are reduced by running the water quickly into the storage tanks. Infiltra- tion losses may be further decreased by compacting the soil surface by rolling, or by spreading a clay blanket before rolling where the surface is sandy but clay is available. Several thousand roaded catchment systems are in operation in Western Australia, some on commercial livestock farms and others providing domestic supplies to villages and small towns. It was estimated that more than 3 500 were in use in 1980. In areas of annual rainfall of 300-400 mm, the proportion of rainfall that can be harvested varies from 24-41 percent, with an average of 35 percent. Water for town supplies may use bitumen to increase run-off but this is usually too expensive for farm use. As a result of this wide experience, detailed design procedures have been published (Frith 1975 and 1977, and Laing 1981b). Designs are also published by the Western Australia Department of Agriculture as Farm Notes 109/84 and 129/84, and a mathematical model for optimizing the combined cost of a dam and roaded catchment is available for graphical solution in Tech. Note 4/76 or computer solution (Frith et al. 1975). The method has also been used on a limited experimental basis in Arizona and Texas where the results are reported to be comparable to those obtained in Western Australia (Hollick 1982).

The flat-batter dam, or spread-bank dam, is another water harvesting method, from the southern coastal area of Western Australia (Plate 6.6). Rainfall is about 400 mm but most falls at low intensities. The terrain is flat or gently sloping sandy soils overlying clay and, together with the low intensity rainfall, this makes roaded catchments unsuitable.

To construct a flat-batter dam, bulldozers and road graders spread the sandy topsoil to form a circular saucer around the proposed tank. Then the tank is excavated into the clay which is spread to a depth of 0.1 m over the sandy saucer to provide the run-off catchment. The inward slope is about 1 percent at the storage tank increasing to 2 percent at the outer rim. The catchment may be from 50-100 m in diameter. The run-off efficiency is comparable with that of roaded catchments, that is, up to 40 percent. This type of dam can be used on flat sites which would not ordinarily produce any run-off, and another advantage is that they can be sited on the top of any high ground so that the water can be distributed by gravity (Laing et al. 1980).

6.2.3 Chemical Treatments

A wide range of chemicals have been tried in attempts to increase the run-off by sealing the surface, and a thorough review of the methods tried is given by Hollick (1982), from which much of the material in this Section has been drawn. In addition to composition of the chemicals, other factors to be considered are the method and the cost of application, the lasting qualities of treatment, and maintenance to ensure that the treat- ment remains effective. All of these have to be costed, since few of the proposed or tested methods show signs of being financially suitable for agricultural purposes, although some may be justified when providing domestic water supplies. Several comparative performance tests have been carried out in USA (Frasier and Cooley 1979; Fink 1984; Fink and Ehrler 1981).

One of the few methods with a long track record is the use of various types of sprays based on cut-back and emulsified bitumen or asphalt, which have been used in Australia for country water supplies for twenty years. There are problems of discoloration, although the contamination is odourless and tasteless. Care and attention to detail in the preparation and the application, and the need for monitoring and maintenance make it less attractive for agricultural use. Laing and Prout (1975) found that for most agricultural purposes compacted clay was more practical and economic. In a later study (1981b) Laing found that compaction alone yielded 40 percent of run-off, one coat of bitumen on compacted earth gave 65 percent, and two coats of bitumen after compaction gave 89 percent run-off. The choice of practice therefore depends on analysis of the costs and the value of the stored run-off.

Another approach is to use water-repellant chemicals, for example waxes have been studied in detail by Fink in Arizona (1984) and silicones by Plueddemann (1975). Tests of many other hydrophobic chemicals are described by Hollick, but so far none have shown the required properties of being cheap and long lasting.

An alternative way of reducing infiltration is to physically block soil pores on the surface. This can be done by adding bentonite, a very fine colloidal clay used for sealing the floor of leaking reservoirs, or mixing cement into the surface soil to form a soil-cement stable layer. Sodium salts have also been extensively studied, for example Dutt (1981) in the USA, and Karan and Julka (1981) in India. They deflocculate the soil so providing fine particles to seal the surface in the same way as bentonite. There are the usual difficulties of cost and the length of the effect, and they have another disadvantage in that the deflocculated soil is less resistant to erosion by the increased run-off.

6.2.4 Waterproof Membranes

Membranes are also reviewed by Hollick (1982). They suffer the same disadvantages of chemical additives by being expensive in relation to their life. An approach which shows some promise, because nine experimental catchments gave a reasonable performance over twelve years (Myers and Frasier 1974), is the combination of a mat to give tensile strength with a sprayed emulsion to make it waterproof. Myers used fibreglass for the mat and several applications of asphalt, but a number of other combinations have been tried.

Flexible membranes have also been used, particularly the polyethylenes and polyvinyls, and the synthetic rubbers, acrylic and butyl. The main problems with the plastics are that they tear easily and degrade quickly; butyl rubber suffers less from these problems but is too expensive for most uses.

Rigid materials are also generally too expensive. Concrete mem- branes have been used and have a reasonably long life but they are very expensive and the run-off is suprisingly low at from 56-80 percent, probably because of the surface depression storage with subsequent evaporation, and the losses from cracks. Corrugated iron sheets have also been used, and do give higher run-off coefficients but the cost limits their application to collecting small amounts for domestic use such as from house roofs, or in Australia where they are the only effective cover which can be applied to the surface of shrinking and swelling black clays. A cheaper material is roofing felt which can be laid in strips with the edges sealed together. There have been a few applications in the USA but the material is too easily damaged and too short-lived to justify the cost.

There have been several comprehensive and detailed studies of alternative methods, particularly in the USA. For example Chiarella and Beck (1975) studied about fifty catchments in Indian reservations using many different treatments and combinations. They came to the conclusion that success was very variable, and the main problem was failure resulting from lack of maintenance since no-one would accept responsibility on communal lands with shared use. A similar trial with similar results was made by McBride and Shiflet (1975), where again the problem was maintenance of communally-used facilities. Mickelson (1975) also reported tests of performance and durability of various membranes and chemical treatments, and in Pakistan a study was carried out with thirty catchment plots and ten treatments (Sheikh, Shah and Aleem 1982). The consensus of these and other trials suggests that technically there are a number of methods which, depending upon the soil and site conditions, can be made to work on an experimental basis. But the difficulties of cost and maintenance make it unlikely that any of the methods so far tested will be economic except in the particular case where small quantities of water are valuable because they are essential for stock or human consumption.

6.3 WATER STORAGE

The harvesting and use of water by storing it in dams, tanks, and other constructed containers is too big a subject for detailed treatment here. This section reviews the principles of the more common types of storage, and for more detail readers are referred to manuals listed with the references.

6.3.1 Small Earth Dams

Small earth dams are probably the most common type of man-made storage, and on favourable sites are effective and economical. They do require some care in design and construction, and they need regular maintenance, but there is a wealth of information and advice available in books and manuals at all levels (Hudson 1975).

i. Planning earth dams

The first step in planning a dam is to define quite clearly the purpose of the dam, so that design requirements can be identified. For example, if a cattle-watering point is required in order to make use of grazing in a particular area, this means there is no point in looking for a site outside the defined area. If the water has to be available throughout the year, this means there must be a minimum depth so that water is always available after allowing for evaporation, but, on the other hand, the amount of water required for stock-watering will be much less than for an irrigation storage dam. When the purpose is irrigation this means the required amount of storage is defined, and also the period when the water must be available. When the purpose is clear, the consideration of the other following factors will be easy.

a. Site location

This is usually dictated by the purpose, as in the above example of providing water in a particular grazing area. When the purpose is irrigation, the location will be dictated by the answers to questions such as:

· "Can the scheme be anywhere in the district, or is there a particular piece of land to be irrigated, and which the dam has to be near?"

· "Is it a gravity scheme where the dam must be above the land? or can we pump from below?"

· "Could we have the dam site upstream and bring the water down in the stream?"

· "If the dam is at some distance from the scheme, would the supply canal have to be lined?"

b. Capacity

In order to plan the capacity of the dam, we need to ask "How much water will be used from the dam, and how much will the demand fluctuate through the year? What is the maximum demand? What are the consequences if the supply dries up? Is a carry-over from one year to the next necessary or desirable? What will the losses be from evaporation and from seepage?"

The shape of the valley influences how much will be stored for a given height of dam wall. The ideal site has a long throwback, that is the surface of the stored water goes a long way back upstream giving a big capacity. It will also have steep valley sides at the dam site so that the earth wall is short. A valley with gently sloping sides means a long and expensive wall. If evaporation is high, it is better to have deep water and a small surface area rather than a large surface area of shallow water. Sites with a good basin storage are often found where the stream passes from a wide valley through a narrow gorge, or where the streambed changes from a flat to a steeper gradient, or at the confluence of two streams so that there are two storage basins for one wall.

c. Soils

The site must have suitable soil conditions for constructing an earth dam. The soil of the basin should not be porous, and the site for the wall should be free from boulders and termite mounds. Suitable soil for constructing the wall should be available nearby. The best kinds of soil for different parts of the wall are discussed under The Embankment or Dam Wall, and Construction Sequence. Certain soils are unsuitable for the construction of dams; these include:

· saline, alkaline, or sodic soils, or any soils with abnormal chemistry;

· peat or other soils high in organic matter;

· heavy clays subject to swelling, shrinking, and cracking;

· light sands;

· soils containing a high proportion of fine silt.

ii. Spillways

Few dams on streams and rivers can be built big enough to store all the run-off, and some provision is required to pass on the surplus flood water after the dam has filled. The spillway is the channel or pipe or waste-weir designed to do this. The cost of small earth dams is greatly increased if spillways have to be built of concrete or similar materials, so a major factor in site selection is looking for sites where the overflow can be safely discharged over grass-covered spillways.

a. Cut spillways

For small conservation dams one solution is open channel spillways cut into the bank at the side of the dam wall (Figure 6.2). These are cheap and effective, but there is a risk of erosion if the water flows too quickly. If it is possible to maintain a good grass cover, this greatly reduces the required size of spillway, but grassed channels do require careful design and construction and may be difficult to establish and maintain in areas of poor rainfall. It is wise to limit the maximum size of catchments for dams with grassed spillways to 500 ha in semi-arid areas.

The most vulnerable point on cut spillways is where the water rejoins the stream. To prevent erosion at this point (A in Figure 6.2) the stream bank must be cut back to a gentle slope and planted with grass like the rest of the spillway. Alternatively, a concrete or masonry drop structure may be provided at this point.

It is undesirable for grass spillways to be constantly saturated. This makes them more vulnerable to erosion because the types of grass which are best for spillways do not like continually wet conditions. So if a stream is likely to have a small constant flow, or one which continues a considerable time after each storm, a trickle flow outlet should be provided. This can take the form of a small brick-lined or concrete channel set in the spillway, or it can be a small diameter pipe going through the dam wall, preferably on the opposite side of the stream from the spillway.

If it is not practical to protect a cut spillway with grass, it may be necessary to protect it, or perhaps the most vulnerable parts, with stone pitching or concrete.

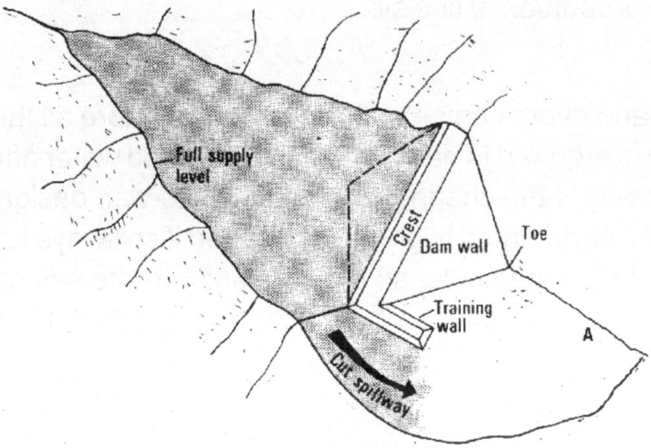

Figure 6.2
Plan of a small earth dam with a cut spillway

b. Natural spillways

These are used where the site conditions make it possible to divert the flood flow into a naturally existing waterway or channel. Figure 6.3. shows an example where the flood is diverted over a crest into the adjacent valley. The crest should be approximately level between A and B so that the flow is spread out, and the slope down to the stream should be gentle and have enough grass cover so that there is no risk of erosion.

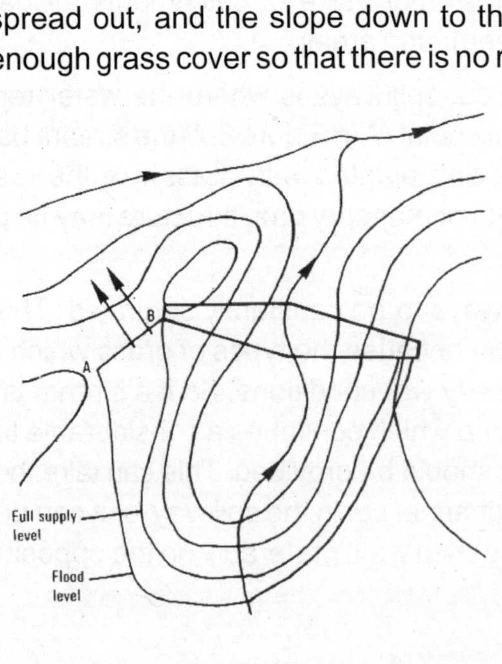

Figure 6.3
A reservoir with a natural spillway into an adjacent watercourse

c. Mechanical spillways

These use pipes to pass the flood through the embankment. It would not be practical to have the pipe large enough to pass the maximum peak flow, so temporary flood storage is provided above the pipe outlet. This fills up during the height of the flood, and discharges through the pipe after the peak flow has passed (Figure 6.4). An emergency spillway is also provided for the exceptional floods. The capacity of mechanical pipe spillways is shown in Table 6.1. Drop inlets are often used at the entry to pipe spillways, partly to ensure a good discharge through the pipe and partly so that the pipe can be buried below the dam wall, as in Figure 6.5.

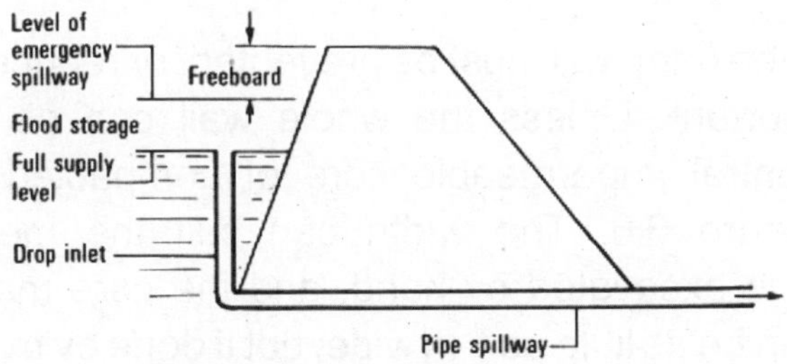

Figure 6.4
A reservoir with a mechanical (pipe) spillway and an emergency spillway

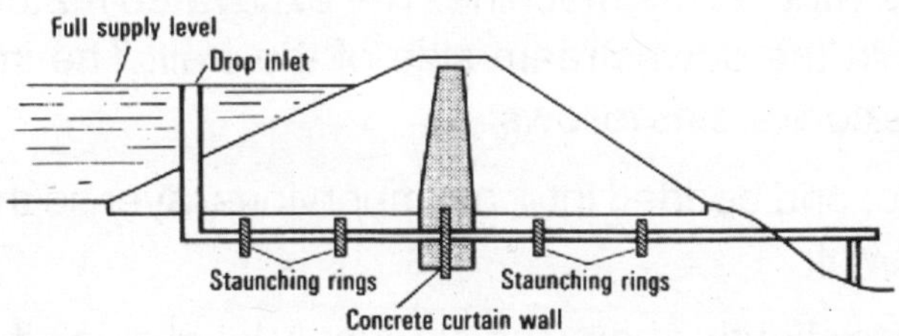

Figure 6.5
Installation of an outlet pipe under a dam wall

TABLE 6.1

CAPACITIES OF MECHANICAL PIPE SPILLWAYS

Diameter of spillway pipe mm	Approximate capacity(1/s)	Diameter of drop-inlet pipe mm
200	5600	200
250	7100	250
300	8500	300
400	11300	500
500	14200	600-750
600	17100	750-1000

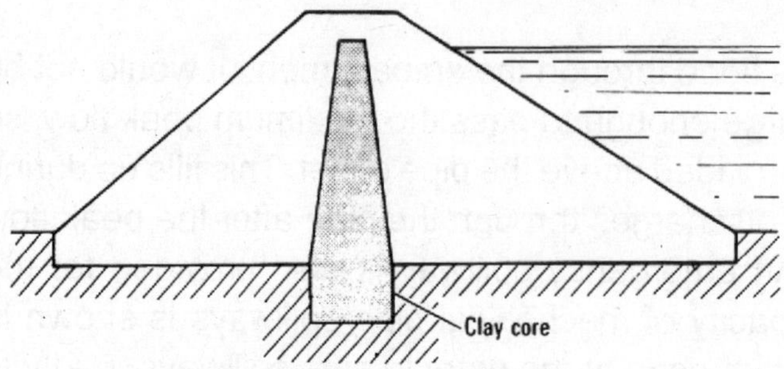

Figure 6.6
An earth dam with an impermeable clay core

iii. The embankment or dam wall

a. Impermeable core

Seepage through or under the dam wall must be prevented, or reduced to the point where it is not important. Unless the whole wall can be built of impervious material, a central impermeable core of compacted clay is required as shown in Figure 6.6. The width can suit the method of construction. If the trench is excavated by hand, and the core material is placed by hand, the core can be as little as 1 m wide, but if done by machine it will have to be at least as wide as the machine. The excavated material from the trench can be used in the downstream side of the wall. The important points in constructing the core are as follows:

· it must be taken down to, and bedded into, an impervious layer so that there is no risk of seepage below it;

· it must be continued up to slightly above the full supply level along the whole length of the wall and at both ends;

· the material used for the core must have enough clay so that it is plastic and workable when wet. For dams which will remain full for most of the time, so the core will not dry out, a pure clay can be used, but for smaller dams which may dry up, a pure clay might shrink and crack, so a sandy clay will be better;

· the material must be properly placed and compacted. The best way is to place the clay in thin layers, add enough water to make it plastic and to 'puddle' it, that is, to compact it by labourers ramming it with poles. Another good method is to drive cattle, oxen, or buffaloes up and down over each layer of clay. When using earth-moving machinery, the soil should be placed by a scraper in thin layers, and compacted when wet by wheeled tractors or by sheepfoot rollers. Placing the soil by bulldozer is not satisfactory, nor is compaction by tracked tractors.

b. Side slopes

The embankment walls must slope inwards at an angle which will be stable at all times. Saturated soil is less stable than dry soil, so the water side of the wall has to be flatter than the downstream side. For most soils the slopes should be not steeper than 3:1 on the upstream side and 2:1 on the downstream. For soils approaching the ideal mixture of grain sizes and which are carefully placed and well compacted, the slopes could be increased to 2:1 upstream and 11/2:1 downstream. Soils which will be difficult to compact should have slopes of 3:1 or 4:1. For small dams the slope does not make much difference to the total costs, so if there is any doubt a flatter slope should be used.

c. Crest width

The top of the dam wall must be wide enough for both the construc- tion equipment and for any traffic if the wall is going to be used as a bridge. Unnecessary width will add to the volume of earth-work required. If the wall will only carry foot traffic and bicycles, 2 m will be enough, but for vehicles a minimum of 3 m should be allowed.

d. Settlement

No matter how carefully the earth is placed and compacted, there will be settlement over the years, and an allowance must be made for this. For average soils with reasonable compaction 10 percent should be added to the finally required height. This means that the newly built wall will be convex along the top as the allowance for settlement will be greater in the middle than at the ends. Embankments built without consolidation, for example those built by bulldozers, are not recommended, but if this method of construction is unavoidable the allowance for settlement should be 20 percent.

iv. Construction sequence

a. Site preparation

The limits of the wall at ground level should be staked out, and all trees, shrubs, and roots removed. Next the grass, grass roots, and topsoil should be dozed off to a depth of about 100 mm. This material should be stockpiled just clear of the site on the down stream side as it will be used at the end of the job to give a dressing of topsoil over the whole wall.

b. Borrow areas

The areas from which soil is to be taken for the embankment should be within the area to be flooded as this increases the capacity, and also avoids bare areas which would be liable to erosion. The borrow area should be near the wall to reduce the haul distance, but not closer than 10 m or seepage under the wall may be increased. The topsoil should be bulldozed off the borrow areas, and stockpiled like the topsoil from the dam wall site.

c. Outlet pipes

It is always difficult to consolidate earth placed around a pipe already in position. The best method is to put the pipe in undisturbed soil under the dam wall, as shown in Figure 6.5. The joints in the pipes must be made carefully and checked for watertightness as leakage would threaten the security of the dam. The staunching rings both hold the pipe in place and also prevent seepage along the pipe. They are cast directly against undisturbed earth, that is without using formwork. If the wall is to have a clay core, the pipe can pass through the core trench, with a concrete curtain wall as shown in Figure 6.5.

d. Building the embankment

The rules for placing the soil of the embankment are the same as for the core, that is, it should be placed in thin layers, and well compacted. Adding water may assist the compaction, but this is a refinement not usually necessary on small dams. Placing soil by bulldozer is always undesirable, because it will not be properly compacted.

e. Finishing works

The embankment should be given a thin cover of the topsoil removed from the site of the embankment and from the borrow pits, and then planted or sown with a suitable grass. The crest and the downstream slope should have a good tough dryland creeping grass, and the upstream side one which will tolerate occasional drowning. The spillway and training wall should also be carefully grassed, and given other protection as required, such as stone pitching.

It is always sound practice to fence off the embankment and spillway, and if cattle will water at the dam it is preferable to fence off the whole dam and to pipe water to a drinking trough below the dam wall.

6.3.2 Small Weirs

Weirs are structures of concrete, brick, or masonry, where the water flows over the top when the storage is full. They may be more suitable than earth dams in some situations, for example:

· when the permanent flow is too big to pass through a spillway;

· when the object is to raise the water level so that part of the flow can be diverted into a canal, but storage is not required;

· when the object is gully control, or to trap sediment.

The main types of weir are shown in Figure 6.7. Gravity weirs depend solely on weight for their stability. A symmetrical section (Figure 6.7a) can be used, but the section in Figure 6.7b gives the same height with less material and . For weirs not built in steps (Figure 6.7c), the back slope is at 2 horizontal to 3 vertical.

The crest width depends on the wall height as follows:
· wall height up to 1 m, crest width 0.3 m;
· wall height 1-1.5 m, crest width 0.4 m;
· wall height 1.5-2 m, crest width 0.5 m.

The weir can be entirely concrete, as in Figure 6.7c, formed by pouring concrete into wooden shuttering. Alternatively, masonry can be used to form the outside layer and act as the shuttering, as in Figure 6.7d. Brickwork can also be used as shuttering as in Figure 6.7e. Another kind of gravity weir is the rock-fill dam (Figure 6.7f), in which a pile of rocks and boulders gives the stability, and a concrete skin makes it watertight.

The large amount of material necessary for gravity weirs can be reduced by slab-and-buttress weirs (Figure 6.7g), or arch weirs (Figure 6.7h) In the slab-and-buttress, the vertical slab is held up against the pressure of water by the buttress, and in the arch weir the pressure of water is resisted by the buttresses at each end. Both these kinds of weir can be built of concrete, brick, or masonry, and can be built up to a height of about lm without design calculations. An engineer should be called in for larger weirs, or in cases where failure would cause serious damage.

i. Construction

The essential requirement for all weirs is a solid rock foundation extending the full width of the stream channel. Building weirs on poor foundations or on a rock bar which only goes halfway across may be possible, but should be left to engineers.

The first step is to clean off thoroughly all dirt and soil, chip off any loose or cracked rock, and if the surface is worn smooth by the water it should be roughened by chipping. To get a good bond between the foundation and the concrete, a grout (cement and water mixed to the consistency of thick cream) is brushed into the rock surface. Brick masonry shuttering is built using a mortar mix of six sand to one cement (by volume) and all joints should be completely filled with mortar. The concrete of a gravity weir does not require great strength, but this should be obtained by the use of an economical mix of good materials, rather than the use of poor materials. The cement should be fresh, and the sand should be clean and free from soil, dirt, and organic matter. River sand is quite suitable if clean; if not, it should be washed. Coarse stone aggregate must be clean and free from soil and dirt. Ideally, it should be a graded mixture of stones from 2 mm to 50 mm. River washed pebbles and gravel are suitable if clean.

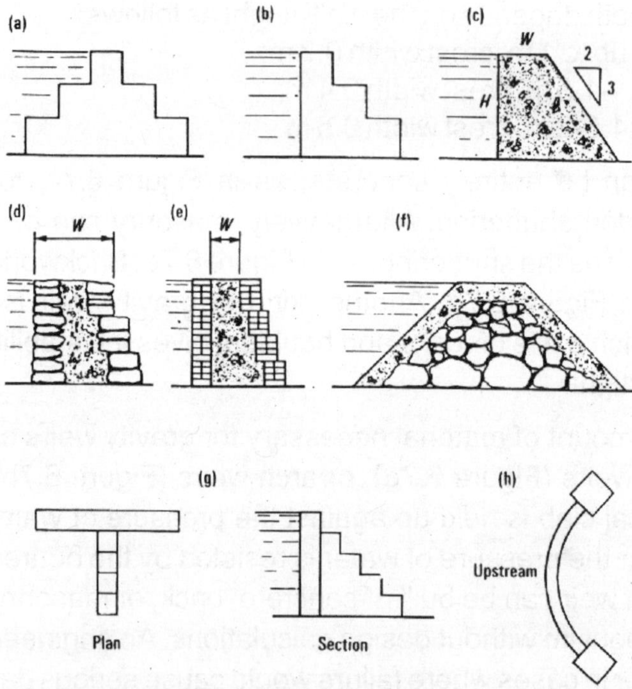

Figure 6.7 Examples of small weirs
(a), (b) & (c) sections through gravity weirs
(d) masonry used as shuttering for a concrete weir
(e) brickwork used as shuttering for a concrete weir
(f) a rock-fill dam with a concrete skin
(g) a slab and buttress weir
(h) an arch weir

A suitable mix is 1 part cement by volume, 3 parts sand, 6 parts stone. The quantities required per cubic metre of concrete are 0.15 m^3 cement, 0.5 m^3 sand, 1 m^3stone, and 70 l of water. The order of mixing is not the same for mechanical mixers and mixing by hand. For mechanical mixers the stone is put in first, then the sand, then the cement, and then the water. For hand mixing, the cement and sand are throughly mixed dry, then the stone, and finally the water.

New concrete should not be allowed to dry out too quickly as this causes shrinkage, cracking, and loss of strength. It should be 'cured' for at least seven days, that is, kept moist by covering with wet sacks. The most common causes of poor quality concrete are:

· using sand or stone contaminated with soil,

· using too much water, and

· not curing properly.

6.3.3 Sand Dams

It is not necessary for water storage to be in the form of free water open to the atmosphere. Sand can store substantial quantities of water in the voids between the particles if they are not filled by smaller particles. A coarse sand of uniform size could store up to 45 percent of its volume as water within the sand. The sand in river beds usually has a mixture of grain sizes but may hold up to 30 percent water, of which about half would drain under gravity. The fact that apparently dry sand river beds contain water at accessible depths is known and exploited by all dwellers of the deserts, both human and animal.

The volume of sand trapped behind a weir will depend on choosing a suitable site, and the storage capacity can be increased over a period of time. If the weir is built progressively in stages of about one metre, the flood will deposit coarse sand behind the weir, but the silt and clay will be carried over the weir by the flood. A second stage can then be added to the weir to trap a second layer and so on progressively to the required height (Plates 6.7 and 6.8). This technique has been used on rivers in the Kalahari desert in Namibia and Botswana in southern Africa, and some of the schemes supply sufficient water for small townships by pumping from a number of linked well points.

There are a number of advantages to storage in this manner:

· once the water level has sunk to 1 m below the sand surface the evaporation loss is negligible;

· the risk of waterborne diseases such as malaria and bilharzia, or contamination by animals, is greatly reduced;

· by installing a pipe through the weir, a supply of clean piped water can be supplied downstream from the weir.

This practice is also used in Latin America where silt trap dams are known as 'trincheras', and in Mexico large traps are encountered called 'atajadizos'.

6.3.4 Off-stream Storage

In some situations it may be impractical to build storage works directly on the stream, and in such cases the answer may be off-stream storage, that is, a storage reservoir which is filled by diverting water into it or by pumping. Off-stream storage is usually more expensive because it loses the advantage of the storage capacity in the stream channel, so it is likely to be used for purposes which justify the higher cost, such as stock watering or domestic use. For cattle watering, the advantage of off-stream storage is that it can be sited exactly where it is required - in a particular paddock, or near a borehole. A special form of off-stream tank storage with its own built-in catchment is the flat batter dam described in Section 6.2.2.

The cost of the water is mainly influenced by the ratio of the volume of storage capacity divided by the volume of excavated earth, known as the storage/excavation ratio or S/E ratio. Just digging a hole and storing water in it would give an S/E ratio of 1.0, but this can usually be increased by using the excavated soil to form a surrounding bank to contain water above the original ground level. A naturally occurring depression may also lead to a good S/E ratio. Several shapes have been worked out which give the best S/E ratio for different conditions. When choosing the best size and shape it is necessary to consider the slope of the ground, the kind of water supply, and the use for the water.

Gently sloping ground from 1 to 4 percent will give better S/E ratios than flat ground, and filling with flood run-off by gravity will be cheaper than pumping from a stream or borehole. In low-rainfall areas, combined systems are common where catch drains are used to increase the catchment area, with a pumping system to top up when rainfall is low (Figure 6.8) In hot, arid climates, an important consideration may be to minimize evaporation by having more depth and less surface area, as discussed in section 6.4.2

Plate 6.7 A simple sand trap weir, Zimbabwe

Plate 6.8 A large sediment trapping dam, Republic of South Africa, using an ingenious method of construction which reduces the amount of concrete required. The shuttering for the arches is made from corrugated roofing iron

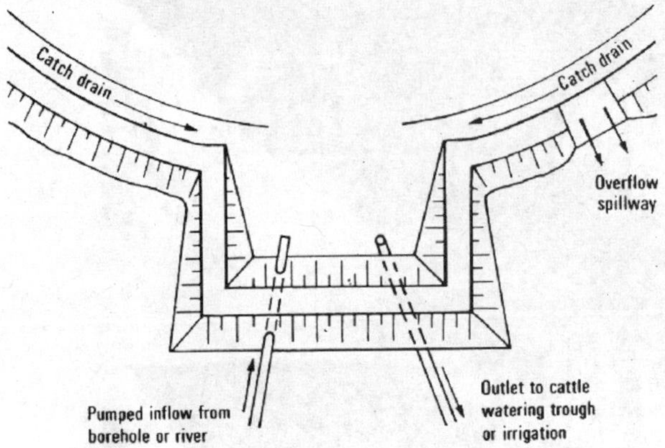

Figure 6.8
Off0stream storage in a tank dam

Earth-moving equipment, which can be used for construction on off-stream storage dams, includes towed scrapers as used for earth dams, and also bulldozers and drag-line excavators. Simpler methods include metal scoops drawn by tractor, oxen, or other draught animals. Prodigious feats of earth-moving have been accomplished in China using wheelbarrows, and in many Asian and African countries by head-carrying earth in baskets.

i. Hafirs and tanks

The storage of water in tank dams is a very ancient practice, and widespread in most semi-arid areas. Throughout Arabic-speaking northern Africa they are called 'hafirs', which may be either water holes in natural depressions, or where the capacity has been increased by excavation, with the spoil used to build a containing bank. Lead-in channels to increase the catchment area are also used. In former times the hafirs were carefully managed, with fences to exclude stock and avoid trampling of the bank, the water being supplied to a drinking point through a pipe. It has been suggested by Sandford (1983) that this style of careful management is only found where the use is limited to a single tribe or group, and that where this no longer applies as a result of social change, the maintenance declines and the hafirs become damaged and silted up. He cites the case in Kenya where, of the hafirs built in north-east Kenya after 1969, most are silted up because no group would maintain them in the face of the policy that they should be open to everybody.

In India also, the use of off-stream scorage is widespread in what are called 'tanks'. It is quite common for them to have sophisticated water control as in the inundation tanks and khadins described in Section 5.3.5, and the tanks are frequently multi-purpose, being for combinations of domestic use, stock consumption, fish production, and irrigation. Deep water pools are sometimes provided within the tank so that fish can survive between rainy seasons, with no fishing permitted in these sanctuary pools.

ii. Rectangular tanks

This is the best shape for gently sloping land and can give S/E ratios of 1.5 to 2.5 (Figure 6.8). Catch drains pick up surface run-off, and when the reservoir is full any surplus is spilled safely through overflows in the catch drains. These are also called tank-dams. Small sizes for stock watering can be built economically using either a drag-line excavator, or a small farm tractor and scoop, or by hand.

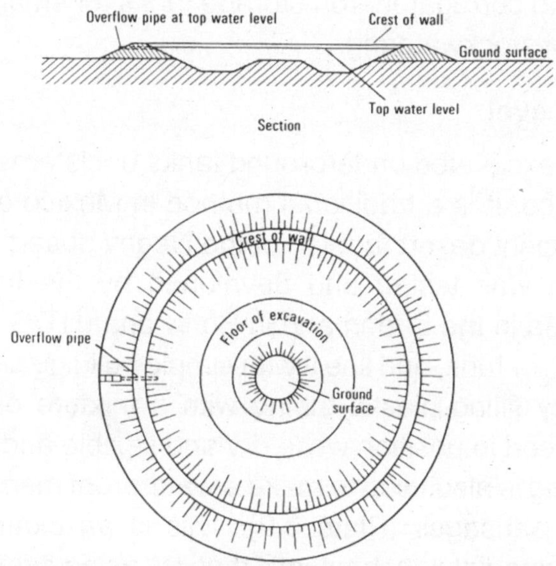

Figure 6.9
Off-stream storage in a ring dam (from Burton 1965)

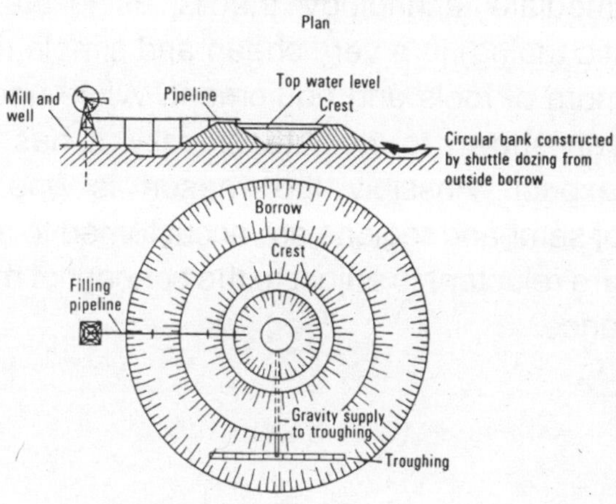

Figure 6.10
Off-stream storage in a turkey's nest dam (from Burton 1965)

iii. Ring tanks

On flat, or nearly flat ground, better S/E ratios are obtained from round tanks rather than square tanks (Figure 6.9). Because of the geometry of this shape, the S/E ratio increases with size, being about 1.5 for a 50 m diameter tank holding about 35 000 m^3 to 4.5 for 200 m diameter holding 200 000 m^3.

iv. Turkeys nest tank

· if the surface soil is relatively impermeable, but overlies a more porous material, or

· where it is convenient to be able to distribute the water by gravity.

The S/E ratio is poor, from 0.2 at 200 m^3 to 0.4 at about 5000 m^3, and the main use is as an alternative to corrugated-iron storage tanks for small quantities of pumped water used for stockwatering.

6.3.5 Storage below Ground Level

The storage of water in excavated underground tanks or cisterns is another ancient and widespread practice. It is a traditional method in Mexico and was an essential component of the ancient desert farming of the Negev desert (Evanari et al. 1982). A modern variation was tested and developed by the Intermediate Technology Development Group in the Sudan and in Botswana (ITDG 1969). The principle is to excavate a cistern or tank, and line it with simple building blocks called 'sausages' which are made by filling plastic tubing with a mixture of sand and cement. The sausages are placed in position while dry and flexible and later set to form a rigid wall. Plastic sheeting is also used to make a waterproof membrane liner on the outside of the plastic sausages (Plate 6.9). This is an example of the successful application of intermediate technology, that is, using the product of advanced technology (the plastic tubing) in a very cheap and simple construction method which requires a minimum of tools and equipment. While undoubtedly a brilliant concept which has been shown to work technically, it has not spread dramatically, as one might expect. Possibly the reason is one commonly encountered; that the peoples of semi-arid regions are accustomed to a precarious way of life, often nomadic, and are reluctant to adjust to the concept of methods and systems which require maintenance.

6.4 LOSSES OF STORED WATER

The main causes of loss of stored water are seepage through a leaking basin or dam wall, and evaporation from the surface. Many methods have been developed for controlling both, but few are economically attractive. The subject has been extensively studied in the USA and in Western Australia, with good reviews of the state-of-the-art by Laing (1975) and Hollick (1982).

6.4.1 Seepage Losses

Losses from seepage can, to some extent, be reduced by site selection, avoiding sands and gravels. The Australian experience suggests that it is often cheaper to build a new storage tank on another site than to try to correct seepage in an existing tank (Laing 1975).

A cheap method which sometimes works is to increase the compaction of the reservoir basin by working it while moist, either by driving wheeled tractors round in the basin, or herding livestock. Sheep in Australia and cattle in Africa have been used to puddle the soil surface in the same way as compacting the clay core of an earth dam (Section 6.3.l.iv). Part of this effect may be the result of droppings from the animals, and the practice of carting in additional quantities of manure to be worked into the surface is reported from South Africa and Missouri.

Other chemical additives are used. One type relies on defloccula- tion to disperse the aggregates and seal the pores. Sodium salts or polyphosphates are used. A colloidal clay marketed under the trade name 'Bentonite' has a flocculating effect from its sodium salt content, and also disperses to fine particles which have a sealing effect. Chemical additives are usually only economic if cheap local supplies are available. The amount required varies according to the soil and should be determined from soil tests for example a well-graded soil will require about 5 kg of Bentonite per m^2 up to 20 kg on sandy soils.

The surface membranes, discussed in Section 6.3, for improving run-off from catchments are all possible solutions to reducing seepage from dams, with the same general conclusion. Thin cheap membranes like the polyethylenes are not sufficiently robust or long lasting, while the stronger and more durable materials like butyl rubber are too expensive.

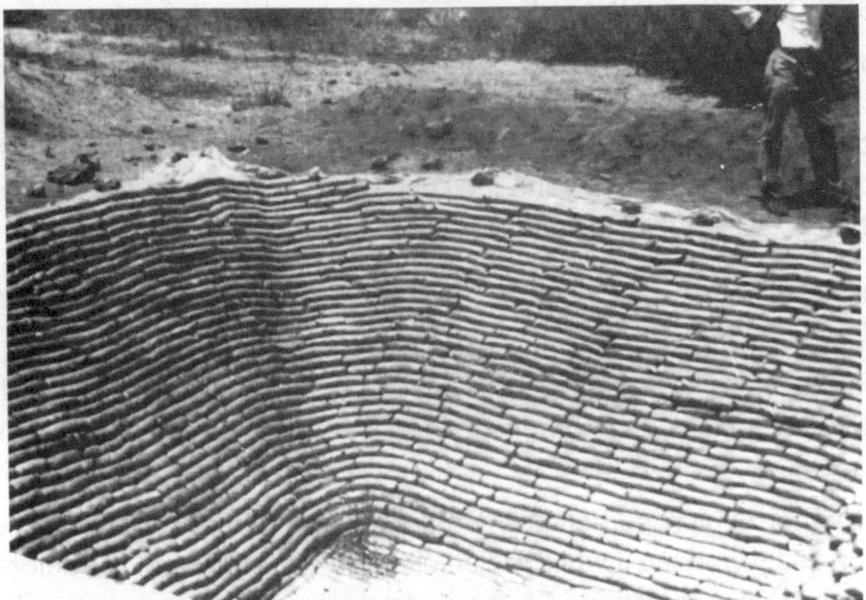

Plate 6.9 Two examples of the 'sausage' method of constructing water storage tanks
(Intermediate Technology Development Group)

6.4.2 Evaporation Losses

To some extent evaporation losses can be reduced by management. If the surface area can be reduced by increasing the depth, this will both reduce the evaporating surface and also lower the incoming radiation and the heating effect. The surface area can also be reduced by storing the water in a compartmented reservoir, and pumping the water from one compartment to another as the water is used, so that there are some full compartments and some empty, instead of a single shallow sheet when the reservoir is partly used (Cluff 1981). Reviews of the alternative methods of supressing evaporation in Arizona and in Australia arrive at different views of the probability of it being worth while. In Australia, Laing and Pepper (1976) conclude that the cheapest solution is to increase the volume of storage to allow for evaporation. On the other hand, Cooley (1975) felt that some methods could be economic, particularly floating covers and the provision of shade.

Shading the water surface can substantially reduce the evaporation. Crow and Manges (1967) showed that a plastic mesh which gave only 6 percent shade reduced the evaporation by 26 percent, while a mesh which gave a 47 percent shade reduced evaporation by 44 percent. However, suspending nets or branches over the water surface is expensive, and no floating mesh has yet been successful.

The method which was thought to have great promise when it was introduced in the nineteen fifties was the use of chemicals which have the ability to form a film on the water surface which is only one molecule thick, but substantially reduces the evaporation. One such chemical was cetyl alcohol. The problem is that the film is easily broken and dispersed by wind and wave action, and the method is little practised today. Another approach has been to use a more robust surface film and possibilities tested include floating covers of expanded polystyrene or rubber (Dedrick et al. 1973), and other floating materials such as blocks of lightweight concrete or floating plastic spheres. All of these can be successfully demonstrated on small experimental tanks and ponds, but have not been adopted for general use.

6.5 DEVELOPING GROUNDWATER

Storage of water within the soil profile was discussed in Chapter 5, and storage using structures and in sand rivers was discussed in Section 6.3. This section is about deep storage and extraction.

6.5.1 Groundwater Recharge

The possible sources of water for recharge are spate flows in streams and rivers, run-off collected by furrows or channels, and run-off stored in tanks or dams.

The conditions required for recharge to be practical are:

· adequate surface infiltration for the water to be absorbed, and permeability so that it can move below the surface;

· good water storage capacity. For rock this is defined as the specific yield, or the volume of water which drains from the rock under gravity. It is expressed as a percentage of the total rock volume, and is similar to the porosity of soils. In rocks it depends on the amount of weathering, joints, cracks, and fissures;

· good hydraulic conductivity is required: the rate at which water can be abstracted from or recharged into the rock depends upon its hydraulic conductivity.

Two problems which may occur with recharge are salinity and silting. Storm run-off and flood water in streams and rivers are usually low in salts because of the dilution, although there are cases of semi-arid saline soils where even the flood run-off is saline. The more common problem of salinity is where raising the water table brings up saline groundwater to within the reach of wells or plant roots.

Silting may be a problem because the flash run-off in semi-arid areas usually carries a heavy sediment load which is likely to block the soil surface and reduce infiltration. Siltation tanks are commonly used to reduce this problem, for example, in the systems of recharging groundwater in the San Joaquin Valley in California.

i. Methods of recharge

The methods of water spreading discussed in Section 5.3 may be used also for groundwater recharge, either by carrying run-off to an infiltration zone, or it may happen almost accidently by seepage through the bottom of canal. Accidental seepage from canals may be beneficial where a rise in the water table is helpful, but it can also lead to undesirable results by causing waterlogging or salinity problems.

If there are aquifers, such as gravel beds, at moderate depth below the soil surface the basin method may be employed. This consists of excavating basins to a depth which extends into the aquifer, so that the basin may be filled from a canal and then left to seep into the aquifer, and then filled again. The inundation tanks or khadins described in Section 5.3 are primarily intended for wetting the soil profile in the floor of the tank, but they have a subsidiary recharge effect, which improves the yield of shallow wells downstream from the tank. When the basins are specially built for groundwater recharge, it will be necessary to avoid them being choked by sediment in the water. Silt-trap tanks may be used, or another method is periodically to remove a shallow layer from the soil surface in the basin.

Another variation, when the aquifer is too far below the surface for the basin system, is to discharge the water into shafts or wells excavated into the aquifer. In this case it is more difficult to correct the effects of choking by sediment, although it can be done by back-wash pumping. A better solution is to use water which is free from sediment.

6.5.2 Groundwater Extraction

We referred in Section 2.2 to the danger of ecological damage resulting from the over exploitation of groundwater resources, particularly fossil water, i.e. water of ancient origin which cannot be replaced. The risk of this happening in arid and semi-arid areas is now increased by the effectiveness of the methods which are available for locating underground water. At the largest scale, satellite imagery is increasingly useful for interpreting the geomorphology, and identifying general areas likely to be suitable for exploration. At the next scale, aerial photographs allow a more detailed study of the hydrogeology, and finally the geophysical survey for particular sites nowadays uses sophisticated techniques including seismic, electromagnetic, and resistivity. All these techniques have been greatly improved in recent years as a result of increased interest in oil exploration.

The exploitation of groundwater for the emergency relief of drought may offer temporary relief while the search for a long-term solution continues. An example of this comes from Zimbabwe. In an emergency programme, 282 small boreholes equipped with hand pumps were installed in six months, and another 40 sites located for hand-dug wells. Electromag- netic and resistivity methods were used for siting, with computer analysis in the field, and this gave a success rate of 90 percent (Houston 1985).

Shallow groundwater may be exploited in the same way as sand rivers (Section 6.3.4). Gravel beds in flood plains or river banks are a useful source, and there is the probability of recharge using flood water.

Shallow wells are a very common way of using groundwater. In India a common design is a stone-lined circular well up to 10 or 15 metres in diameter and similar depth, or rectangular wells of similar capacity. This large capacity is required because both the clay soil and the underlying rock have a low hydraulic conductivity. The well is emptied each day by domestic use, stock watering, and small-scale irrigation, and isreplenished overnight. This traditional method has been successfully operated for centuries when the water was lifted by hand or ox-power, but in some districts the widespread introduction of first diesel-engined pumps and more recently electic motors has meant that the water can be removed much more quickly. This has led to searches for increased capacity and delivery. In the Deccan Plateau, a common device is to drill small bores (30-50 mm) radially and horizontally outwards in an attempt to tap more faults and fissures in the rock. The same technique of increasing the yield of wells is used in Iran where the horizontal collectors are shafts or galleries driven by hand labour. Simple water-lifting devices are reviewed by Kennedy and Rogers (1985).

Another type of horizontal drilling is used in the western United States of America (Welchert and Freeman 1973) to supply small quantities of water for watering stock. These horizontal wells are also called 'cased springs' because they seek to develop existing springs, as shown in Figure 6.11(a) or to release contained water as in Figure 6.11(b). Using special drilling equipment a hole is drilled horizontally and a metal casing of 50 mm diameter is inserted. The pipe is sealed into place and fitted with a valve so there is no danger of contamination. The hole is drilled at a slight downward angle so that the pipe remains constantly filled with water and flows out under the head in the aquifer.

This is a modern development of the age-old system of driving a horizontal shaft into the side of a hill until it reaches the water table, so that water then flows out through the tunnel under gravity (Figure 6.12). This method is widespread throughout northern Africa and western Asia. In Afghanistan and Pakistan they are called 'karezes', in north Africa 'foggaras', in the United Arab Emirates 'falaj', and in Iran 'qanats'. It is estimated that in Iran there are 40 000 qanats with a total length of 270 000 km and supplying 35 percent of the water supplies of the country, mainly in areas with mean annual rainfall of 150-250 mm. The entire water supply for the 2 million population of the city of Teheran was supplied from qanats until recently, when a storage reservoir was built. The construction of qanats was extremely arduous and dangerous, and few are excavated nowadays, although hundreds of thousands are in everyday operation. It seems possible that using modern techniques of tunnelling, with acceptable safety standards, it should be possible to develop a present-day version of this ancient and successful method.

(a)

(b)

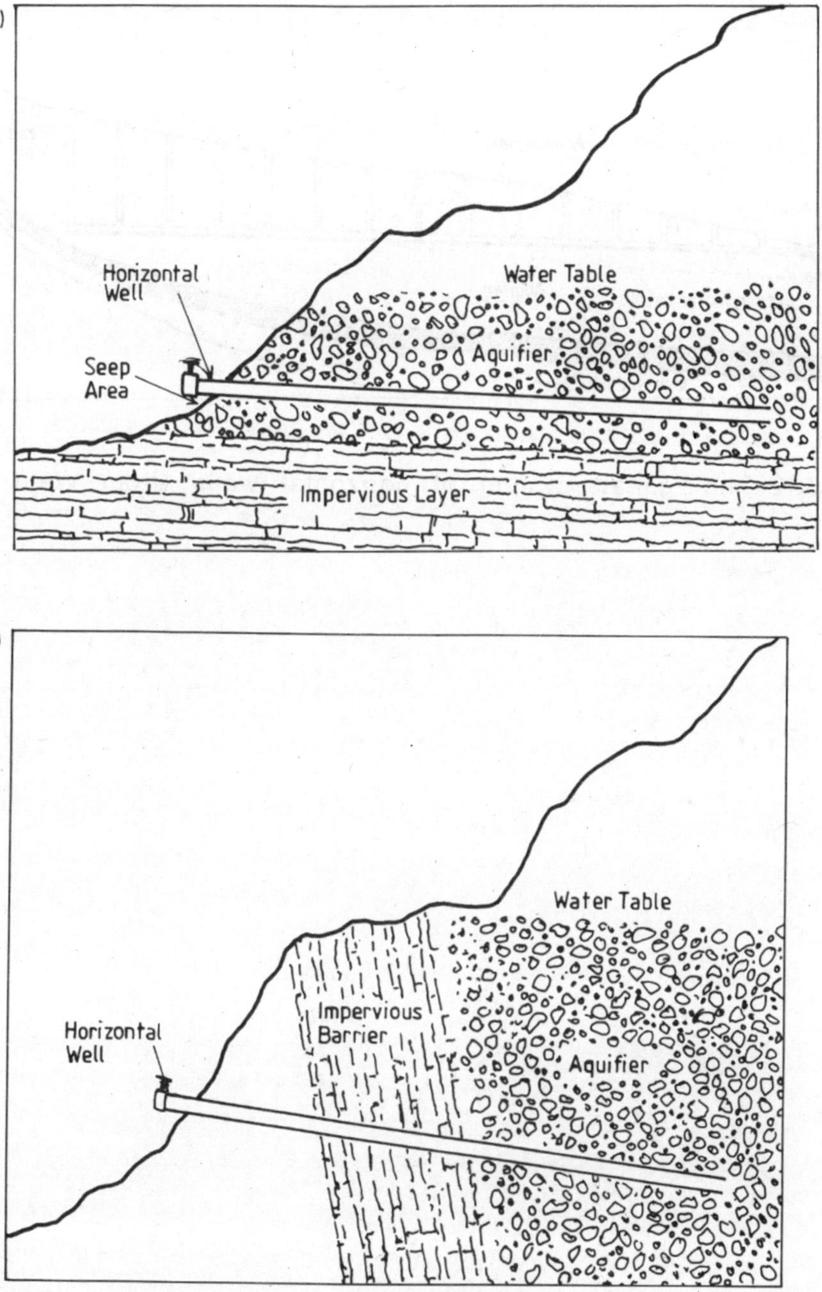

Figure 6.11 Horizontal wells

(a) drilled into an aquifer which is detected by the better vegetation growing at the seepage area

(b) well drilled into a contained aquifer

(after Welchert and Freeman 1973)

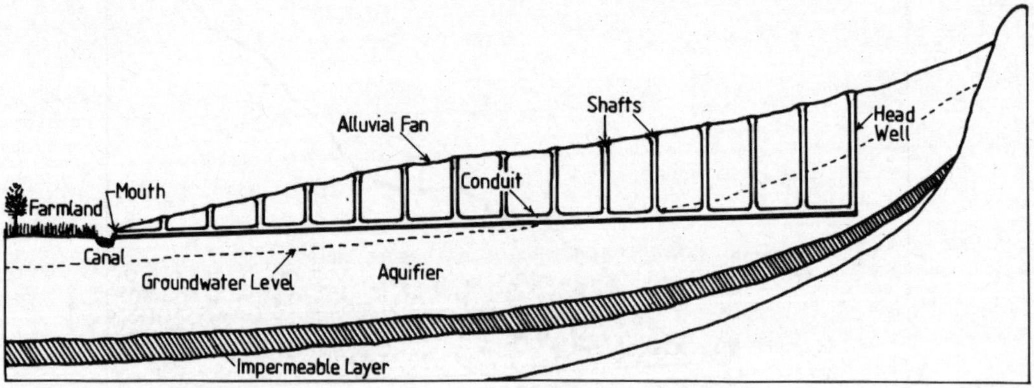

Figure 6.12 The 'ganat' system of horizontal wells (from Wulff 1968)

7. APPLICATIONS OF WATER CONSERVATION

7.1 GRAZING LAND

7.1.1 The Issues

It is not appropriate to attempt to discuss the whole subject of the management of grazing land, but only to look briefly at some applications of water conservation and management on grazing land. Reviews of the experience of range land management in Australia are given by Harrington et al. (1984), and in the USA by Valentine (1971). There is widespread concern about the degradation of grazing lands, particularly in the semi-arid areas of Africa (Plate 7.1). There is also anxiety that in some cases the increased degradation is the result of well-meaning development policies, an argument very convincingly presented by Ormerod (1978). The evidence of satellite imagery proves conclusively that the degradation of the Sahel and south west Africa is man-made and caused by overgrazing, and

Plate 7.1 Reduced vegetation from overgrazing in the semi-arid region of Zimbabwe

Plate 7.2 Satellite imagery showing overgrazing, Namibia. The dark areas are where overgrazing is controlled, at upper left on a well-managed commercial cattle ranch, and at lower right in the protected Etosha National Game Park
(Remote Sensing Centre, FAO)

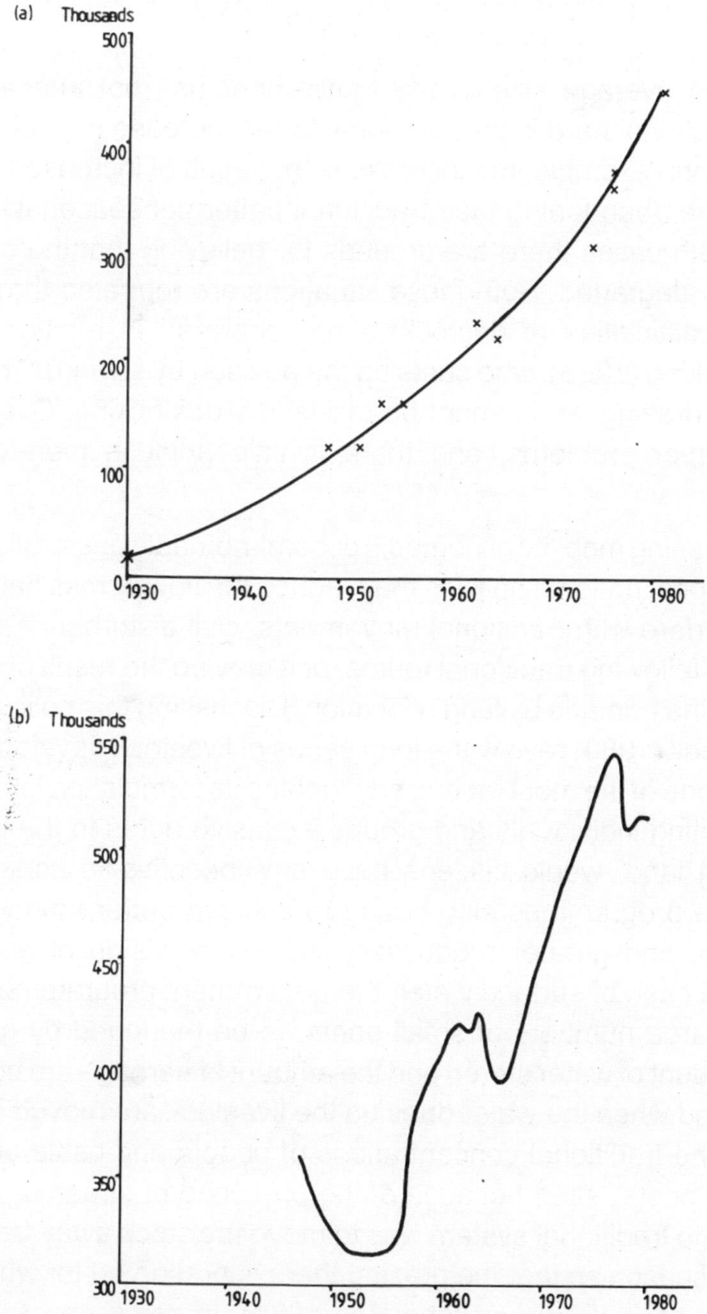

Figure 7.1 Changes in the cattle population
(a) Mambilla Plateau, Nigeria, 1930-1980
(b) Swaziland, 1947-1980

One problem is the increased pressure on the land which arises from the increasing size of most national herds. Two examples are shown in Figure 7.1, from the Mambilla Plateau on the border of Nigeria with Cameroon and from Swaziland. In Swaziland, the average size of the family herd has not increased, and the increase in the national herd is propor- tional to the increase in population (Fowler 1981). In the Nigeria example, the increase is the result of increased settlement by the Falani who are abandoning their traditional pattern of seasonal migration with their herds. In both cases there are grounds for believing that the eco- system is being irreversibly degraded. Both these situations are repeated throughout semi-arid Africa. The difficulties of destocking are analysed in relation to a tribe in Tanzania by Stocking (1985), who sums up the position by saying, "The message is clear: erosion of grazing land cannot be tackled by destocking. Overstocking is a symptom of deeper problems, and there is little point in merely treating the symptoms."

The decreasing mobility of nomadic or semi-nomadic pastoralists arises from a number of reasons; national policies may reduce the flow across national borders, fencing may interfere with traditional movements, civil disturbance may deter the pastoralists from following traditional routes, or it may be the result of a government policy of settling the nomads by land allocation. Discussing this policy in Botswana, M.L. and M.J. Odell (1980) review the long series of livestock development projects and, referring to one of the most recent, say "behind its formulation lay the conviction that only by granting individuals and groups exclusive rights to the use of defined areas of grazing land, would citizens have any incentive to control their stock numbers." These programmes also hoped to achieve better management of the stock and range, and greater productivity, by the provision of additional water resources. In the case of surface water, the government programme sponsors the construction of large numbers of small dams, to be managed by groups of local farmers. The amount of water stored and the amount of herbage are both dependent on the rainfall, and when the water dries up the livestock are moved on to pastures new. However, the traditional concentrations of people and cattle are at what are known as 'cattle posts', sited because of the presence of streams, wells, or other surface water. The traditional system was to move the stock away from the villages during the rainy season so that the grazing there is preserved for when they return during the dry season (Fortmann and Roe 1981). In some cases, the source of supply has been replaced by boreholes which draw water from the underlying Kalahari sands. This has improved the availability of water close to the cattle posts, and reduced the seasonal movement to traditional grazing areas at some distance, with the result that there is severe local overgrazing and degradation close to the population centres.

Two reports of successful group management schemes should be mentioned. From Niger, Pacey (1986) reports the formation in the early nineteen eighties of herders' associations of about 20 households to manage grazing land on a cooperative basis. The other example is a UNDP project in eastern Senegal (UNDP 1984). In an attempt to overcome the problems of communal use of grazing, permanent rights were allocated to units of 100 families. It was hoped that this would lead to:

- grazing management plans to allow the more effective use of grazing land;
- incentives to increase income through increased offtake and more productive herds;
- a framework for investment in water facilities, livestock extension services, training, and credit.

By and large, the programme was considered successful. Between 1977 and 1983 the herd composition improved, mortality declined, and calving rate increased. Both offtake and herd size increased. Between 1980 and 1982 the average income per family from livestock increased 48 percent owing to an increased offtake combined with better prices.

7.1.2 Management of Grazing Areas

i. Water management

Water management is usually the key to increasing the productivity of semi-arid grazing land. There are many other aspects of good management, the use of an appropriate stocking rate, rotational grazing with rest periods, the controlled use of fire, and improving the herbage composition, but water management is usually the most important. In field trials in western Colorado, grazing control alone reduced run-off by 30 percent and sediment yield by 35 percent (Hadley 1977). Changing the vegetation from woody shrubs to grass in Arizona decreased the amount of run-off and increased the carrying capacity, and the authors comment "considering the large amount of semi-arid range land areas of the world, of which the pre-treatment area was typical, such range land conversion can have a tremendous impact on the agricultural and economic output." (Simanton et al. 1977). In a factorial field experiment in Nebraska with a mean annual rainfall of 500 mm, the run-off saved by contour furrowing was up to 30 percent of the rainfall (Dragoun and Kuhlman 1968).

Another example of a close linkage between grazing management and water supplies, similar to the Botswana example in the previous section, comes from Butana region in the Sudan, with a mean annual rainfall of 200 mm. Present day nomadic herders continue a practice thought to have started more than 2000 years ago. Wells are dug to store water during the summer thunder storms. But when the rains start, the herds are moved out from the Atbara river onto the grasslands of the Butana, returning later to use the stored water and the nearby grazing (Hays 1981).

Another variation of using water supplies to manage the grazing comes from Arizona with a mean annual rainfall of 150 mm. A 1500 ha range had three permanent water supplies and two seasonal supplies. Instead of controlling the grazing at high cost by fencing, it was managed by fencing off each water supply, and opening them one at a time. The cattle were encouraged to use the open supply by placing salt blocks there. During the first season the cattle had to be driven away from the closed water but within a year had learned to move to wherever the water was open (Martin and Ward 1970).

ii. Large-scale surface management

A number of methods are used to reduce surface run-off by manipulating the soil surface so as to slow down the run-off or to increase surface depression storage. Since these usually involve catching and holding water temporarily in furrows or depressions, this can be used to assist germination when reseeding to improve the herbage species composition, or to introduce new species. Water spreading by diverting flood flows was discussed in Section 5.3.2. Simple methods of spreading the water may give a distribution which is not sufficiently uniform for cropping, but can improve the grass production at low cost, even if the effect is patchy. This is an old established practice in the semi-arid western states of the USA as reported by Valentine (1971). An example from Australia is shown in Plate 7.3. The productivity of grass can be impaired by the deposition of sediment from floods carrying a high silt load, so sediment traps are sometimes included in the distribution system.

Large spreading banks may be a nuisance if the land is in a rotation with arable crops. In this case a smaller version can be used, as shown in Plate 7.4 from Western Australia. Here wheat alternates with a self-sown pasture of grass and sub-clover. The small pasture furrows are run on a true contour with all the excavated soil spread thinly on the uphill side. The furrows catch and spread small run-offs, and if the run-off exceeds the storage it spills over the downhill edge without causing an erosion problem. Machines have been developed in the USA to make small furrows and apply fertilizer, grass seed, and legume seed in a single pass (Erickson and Currie 1982). In flat lands in Australia, small pasture furrows have been used in rectangular pattern to encourage ponding and infiltration on degraded range land (Cunningham 1974). Contour furrows have been shown to be effective in the USA by Dragoun and Kuhlman (1968) and by Neff and Wright (1981). The effect is sufficient to justify expensive operations for commercial ranching, but they are not appropriate for nomadic or semi-nomadic graziers at subsistence level.

iii. Small-scale surface modifications

It may be possible to substitute inputs of labour for inputs of money or machines, and an example is the semi-circular hoops discussed in Section 5.2.4. In Niger, small basins (cuvettes) are formed by digging small furrows (20 cm deep), with hand-hoes, in a square grid at 10 m intervals. Dung is spread in the basins, and grass seed scattered to give improved grass growth (Gallacher 1974). Another method requiring only labour and local materials is the building of low stone banks, discussed in Section 7.2.2 and shown in Plates 7.5 and 7.6 as 'diguettes'. These may be used for growing fodder or food crops, with the choice depending on the rainfall and the wishes of the local population. In Burkina Faso and in Kenya, projects have started out with the object of improving food for livestock, but have later changed the emphasis towards food production. Another example of changing the objective occurred in the Negev desert where microcatchments were used to direct all the run-off from a basin to the lowest point where a fruit tree could be grown. The nomadic Bedouin felt a greater need for pasture shrubs than fruit. Microcatchments are described and illustrated in Section 7.2.

iv. Implements and machines

Considering tools for improving the surface storage, the principle of holding rainfall in small furrows or depressions, discussed in Section 5.2.2, can also be applied to range land, and several machines have been developed for the purpose. One is a modified disc plough shown in Figure 7.2, where some of the plough discs are removed and those left have a piece cut out of the disc,

Plate 7.3 Spreading floodwater to improve pasture, Australia (Soil Conservation Service of New South Wales)

Plate 7.4 Pasture furrows to spread surface run-off, Austraia

so that the effect is to produce shallow plough furrows in short sections. Using this method in Arizona was shown to be effective in both reducing surface run-off and increasing the density of the grass cover (Tromble 1976). In Australia two implements to achieve a similar effect are the opposed disc pitter (Plate 7.7) and the ripper plus opposed discs (Plate 7.8). The result is shown in Plate 7.9 in the north west of Western Australia, and Plate 7.10 in western New South Wales. Another machine for creating pockets of high infiltration and storage is the tine pitter, which consists of heavy duty tines mounted on a rotating shaft or drum, and a variation, the spiked roller, is shown in Plate 7.11. A machine which improves surface storage by creating a corrugated effect is the land imprinter, developed in Arizona (Dixon 1983), and also manufactured in Australia (Plate 7.12).

7.1.3 Reclamation of Degraded Range Land

Where grazing land has been degraded by overgrazing, the capacity for it to recover when rested is sometimes quite remarkable. Plate 7.13 shows an example from Ethiopia, where the total exclusion of all livestock on one side of the river gave a rapid recovery. Another example comes from India, shown in three photographs in Plate 7.14. Total exclusion was enforced by a watchman, and trees for fodder and fuel were planted in a programme operated by the local community. Again, there was a rapid recovery after only a few seasons.

Where the degradation is exaggerated by drought or unreliable rain- fall, more intensive reclamation using machinery may be necessary. Some of the techniques are demonstrated in Plates 7.7 to 7.12, and 7.15 and 7.16. Two successful projects are reported from Australia; one in Central Australia reported by Keetch (1981), and the other in Northern Territories reported by Fitzgerald (1976).

Another reclamation activity extensively studied in Australia is improving what are known locally as 'scald' areas, where a combination of accumulated salinity and surface crusting leads to an almost barren surface. Many techniques have been thoroughly tested, and one of the more successful is ponding behind low banks to leach out the salinity as reported by Jones (1967) and illustrated in Plate 7.17. Others are ripping and furrowing in 'spirals', Plate 7.18, and sowing grass in pits (Plate 7.19)

Plate 7.5 Marking out semi-circular hoops called 'diguettes' with a shallow excavation, Burkina Faso (OXFAM)

Plate 7.6 Building the low stone wall with lumps of laterite, Burkina Faso (OXFAM)

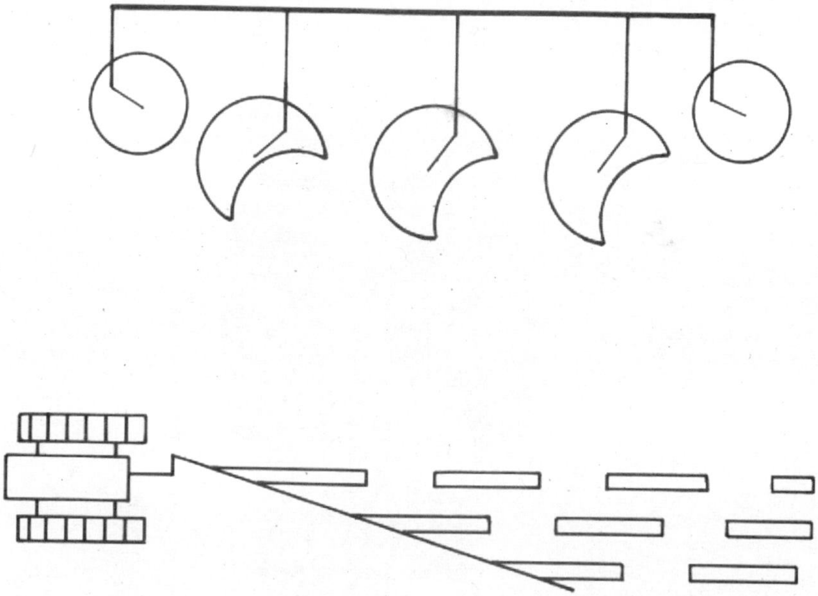

Figure 7.2 A modified disc plough for making pits on range land

Plate 7.7 The opposed disc pitter to improve surface storage. Note the universal coupling between the two pairs (Western Australia Department of Agriculture)

Plate 7.8 Opposed discs with ripper and grass seeder (Western Australia Department of Agriculture)

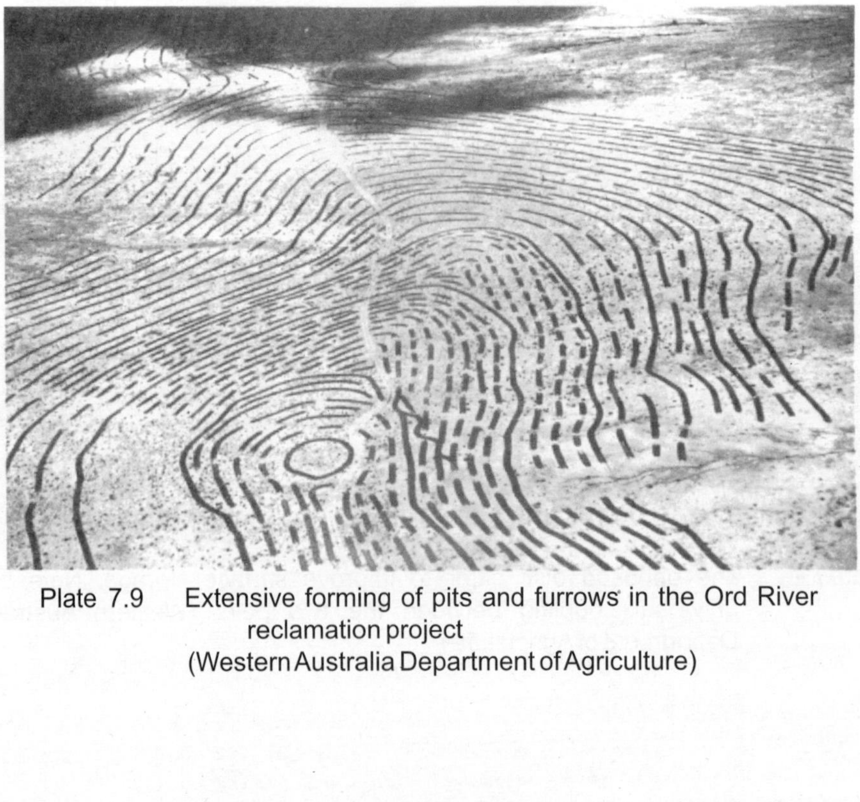

Plate 7.9 Extensive forming of pits and furrows in the Ord River reclamation project (Western Australia Department of Agriculture)

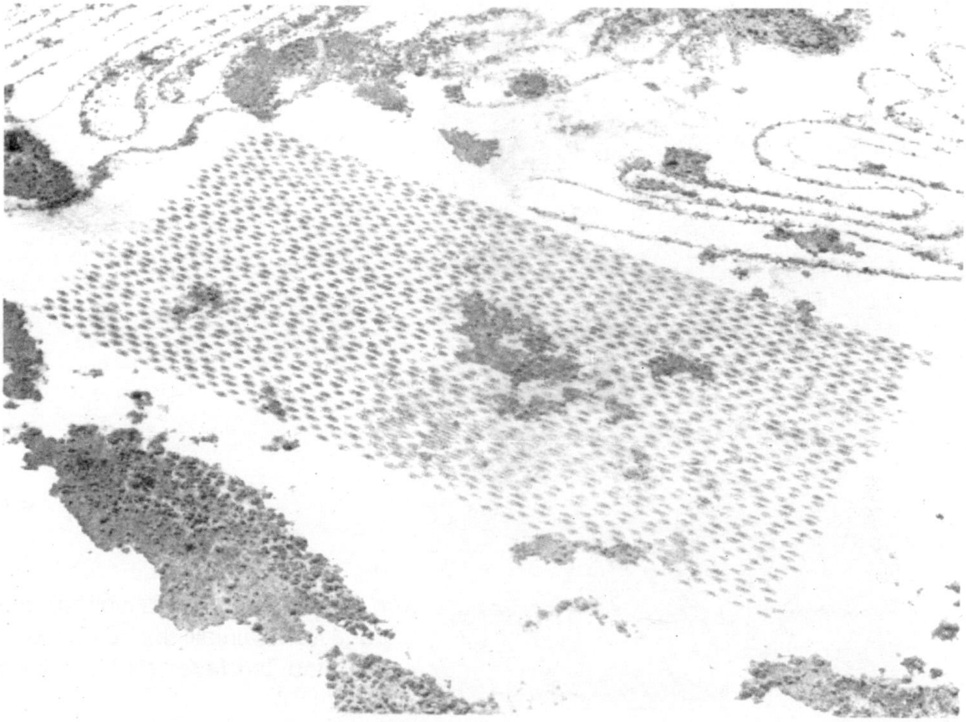

Plate 7.10 Range pitting in the semi-arid western block of New South Wales, Australia (Western Australia Department of Agriculture)

Plate 7.11 A spiked roller for range improvement, which can be filled with water to provide extra weight if necessary (Western Australia Department of Agriculture)

Plate 7.12
The imprinter improves soil surface storage of rainfall by creating a corrugated surface (R.M. Dixon, USDA-ARS)

Plate 7.13 The effect of excluding grazing, Ethiopia, with complete exclusion on the right bank and continued overgrazing on the left bank

Plate 7.14
Restoration of grazing land, India
(T.F. Shaxson)

(a) the communal grazing land is almost denuded by overgrazing

(b) the village council agree to try a policy of total exclusion of grazing and tree planting, and appoint a watchman

(c) the trees prosper and the grass recovers after only two seasons

Plate 7.15 A roller chopper to control undesirable shrubs invading range land
(Soil Conservation Service of New South Wales)

Plate 7.16 Staggered furrows to increase infiltration using the opposed disc plough and seeder shown in Plate 7.8
(Conservation Commission of the Northern Territory, Australia)

Plate 7.17 Ponding banks to leach salinity and encourage re-vegetation on 'scald' (Soil Conservation Service of New South Wales)

Plate 7.18 On level scald areas spirals of furrows are used to encourage regrowth. On soft ground the opposed disc plough of Plate 7.7 may be used alone; on harder ground the ripper shown in Plate 7.8 is necessary (Conservation Commission of the Northern Territory of Australia)

Plate 7.19 Buffel grass (Cenchrus ciliaris) seeded in pits on scald area near Alice Springs. A tine pitter is used, with an eccentric wheel for intermittent lifting. The grass in sown at the same time as pitting (Conservation Commission of the Northern Territory of Australia)

7.2 TREES AND SHRUBS

7.2.1 The Issues

In semi-arid areas we do not find woodland, bush, grazing land, and arable land occurring as separate forms of land use. The landscape is always a mixture of some or all of these, and trees or shrubs will always be a part of the ecosystem. They supply many needs: fuel, browse, construction, food crops and cash crops, and usually a combination of more than one of these.

Depletion of the natural stock of trees and shrubs is one of the indicators of excessive pressure on the land leading to an increasing rate of degradation. The pattern of progressive degradation is described by Brown (1985) based on the model of Newcombe after studies in Ethiopia. In the first stage, natural forest retreats before the plough, but wood supplies are adequate and the gradual decline in crop land fertility is not noticeable. As populations grow, markets appear for wood which is cut and sold by peasant families who turn to crop residues and animal dung for their own fuel. Depletion of the forest accelerates; so does the decline of crop land fertility, and the rate of soil erosion.

As the nearby trees disappear, dung and crop residues are increasingly traded in markets where formerly only wood was sold. Eventually cow dung becomes the main fuel source in villages and thus the main cash crop from nearby farms. Rural families use crop residues for cooking, and as forage for their livestock which the grazing land can no longer support. Crop failure increases on the eroded and depleted soil, and droughts which could formerly have been tolerated, now become a disaster. In the final stage, a sequence of poor rainfall seasons causes the collapse of the system. Families cannot feed themselves or their livestock, and there is a massive exodus. As Newcombe says "this transition from the first to the final stage is in process right across Ethiopia and has reached the terminal phase in parts of Tigre and Eritrea." Similar patterns are unfolding in many other semi-arid areas; in the rest of semi-arid eastern and southern Africa, in the Sahel, in India, in north-east Brazil, and so on.

In many countries the increasing scarcity of fuelwood is becoming as critical as the scarcity of food (Eckholm 1976, Chapter 6). In India it has been estimated by the National Committee on Agriculture that the present day fuelwood need of India is 250 million m^3, plus about 60 to 80 million tons of dry cow dung, and the deficit is expected to reach 100 million m^3 by 1990.

In this context, the need to maintain or expand the production of trees and shrubs through soil and water conservation methods and programmes is clearly a major challenge in the semi-arid regions.

7.2.2 Methods and Techniques

The usual techniques for afforestation are seldom appropriate in semi-arid areas. The unreliability of the rainfall makes it difficult to get the trees or shrubs through the initial stages of growth, even if the annual rainfall is sufficient to maintain mature plants of the same species. The soils are frequently poor and shallow, with low moisture storage capacity. Even if these physical problems can be overcome there remains the difficulty of protecting the young plants. In pastoral economies there is little use of fencing, except occasionally to protect food crops, and the usual pattern is either completely free grazing or light control with rather casual herding. Where total exclusion of stock is possible, like the example described in Section 7.1.3, the protection of the grass can be extended to trees and shrubs.

Several methods have been developed to increase the amount of water vailable for tree production using the rainfall multiplier approach escribed in Chapter 5. Microcatchments, or 'negarin', were developed as art of the Israeli studies of run-off farming in the Negev desert (Shanan and Tadmor 1979). A typical design is to use a basin of 250 m^2 as the catchment area and surrounded by a low earth bank. At the lowest point within the basin, a deeper basin is excavated about 3.5 m x 3.5 m, in which one tree seedling is planted (see Figure 7.3). The system has been successfully used for the establishment of almond trees and fruit trees as shown in Plate 7.20. In Kenya, in an area of rainfall of 300-600 mm, a slightly different method was used. Planting holes were dug 600 x 600 mm and 300 mm deep, with a collecting drain to form two arms of a V leading run-off into the planting hole. In heavy rain run-off surplus to the capacity of the hole spills over at the end of the collecting drains. The recommended length of the collecting drains is 3-5 m and the spacing is 5 m x 5 m or 400 trees per ha (Barrow 1983). Other variations are small half-moons (lunettes) as in Plate 7.21, and lines of small semi-circles or Vs as in Figure 7.4. In very low rainfall areas, the yield of micro- catchments can be enhanced by treating the surface to increase the run-off and this was demonstrated by Rawitz and Hillel (1975), who found that it was necessary to treat the surface of deep loess sandy loam by compacting and spraying with diluted fuel oil. With a catchment ratio of 3:1 and rainfall of 200-250 mm this gave an effective rainfall at the planting site of 600 mm which was satisfactory for almond orchards. The cost of preparing and maintaining microcatchments means that it will only be economic for high-yielding crops such as fruit or nuts, unless a simple low-cost method can be used as in the Kenya example (Boers 1983). The imprinting method mentioned in Section 7.1.2 for establishing grass is also used for establishing seedlings (Anderson 1986).

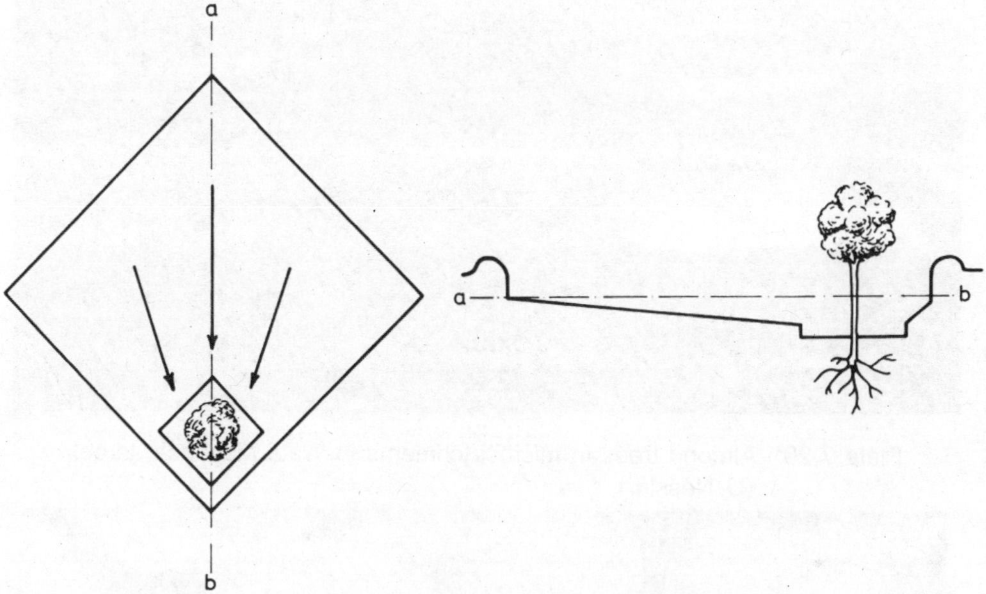

Figure 7.3 Plan and cross-secftion of negarin plot; the arrows indicate the direction of run-off flow (Evenari et al. 1982

The same principle can be used by planting trees or shrubs in lines where increased water is retained by some type of contour furrow. In Tunisia this has been practised using very simple stone cordons, which are triangular loose piles of stones made on the contour and just downhill from where lines have been drawn with a deep ripper and a heavy-duty tractor. The combination of loosening the surface and breaking through a calcareous pan improves the infiltration and the water is ponded behind the cordon as shown in Figure 7.5 (Viertmann 1984). For the establishment of shelter belts in Iran, terraces were built on the contour, 2 m wide and at a 5 m interval on 30 degree slopes. The survival rate of seedlings was substantially increased, particularly if the catchment area was sprayed with asphalt (Kowsar 1982).

Plate 7.20 Almond trees in microcatchments in Wadi Mashesh, Israel (U. Nessler)

Plate 7.21 'Lunettes' or half-moons for tree planting in Niger (R. Gallacher)

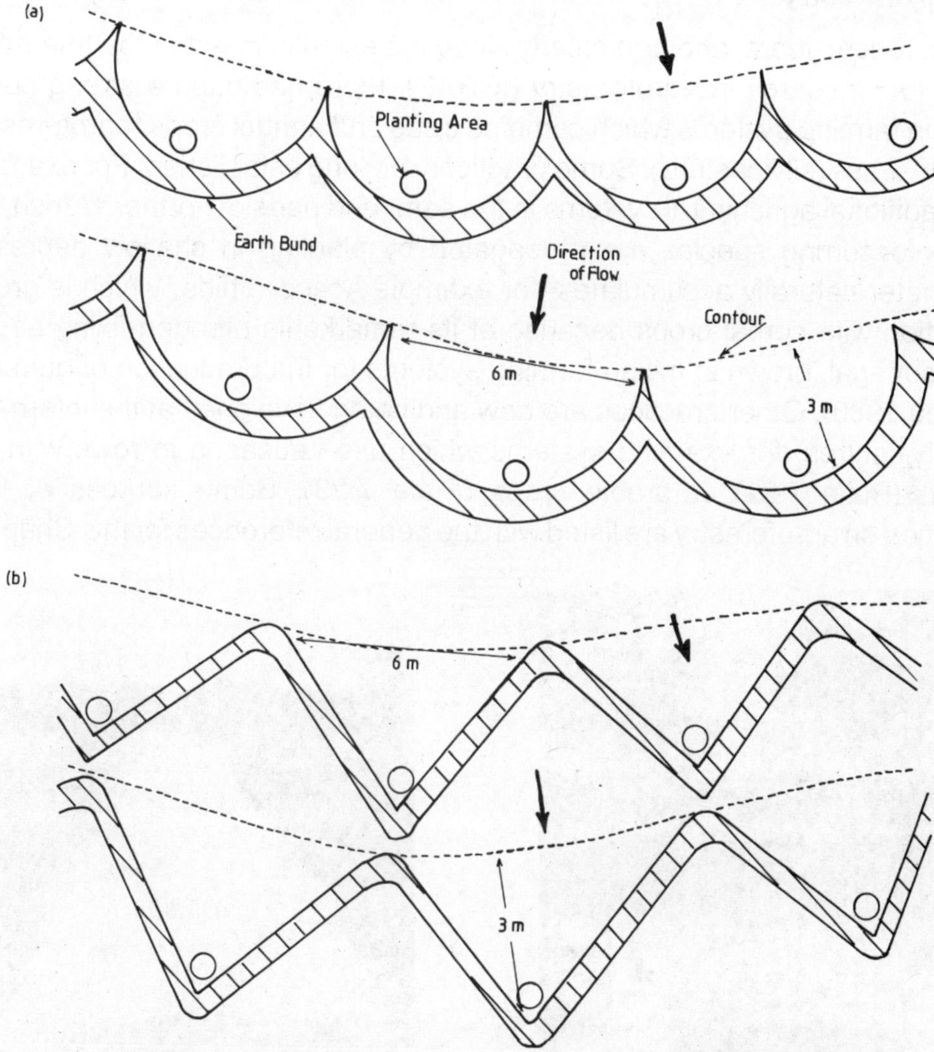

Figure 7.4 Small microcatchments for planting trees (Baringo Pilot Semi-arid Area Project)

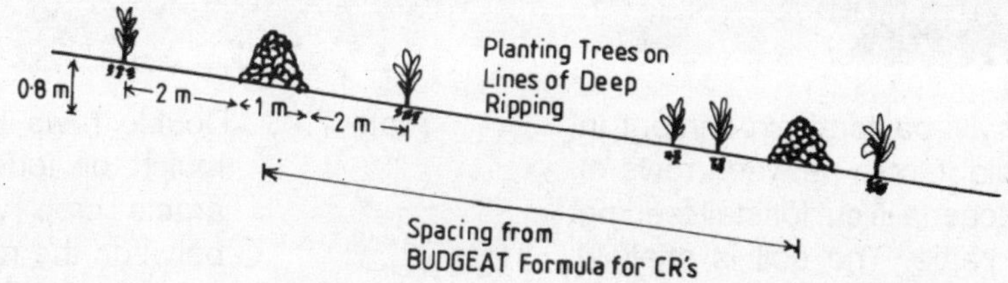

Figure 7.5 Planting trees long lines of deep ripping next to stone lines ('cordons') in Niger

7.2.3 Agroforestry

In recent years, and particularly since the establishment of the International Council for Research in Agroforestry (ICRAF), there has been a strong surge of interest in farming systems which combine trees and annual crops (agroforestry) or trees and grass (silvipasture). Some practices are long established. For example, in many traditional agricultural systems in the semi-arid parts of northern Africa, some naturally occurring species are propagated by planting in shallow depressions where water naturally accumulates, for example Acacia albida, which is grown in association with cereal crops because of its remarkable nitrogen-fixing ability, or Acacia senegal, grown in mixed farming systems for the production of gum arabic (Houerou 1980). Other practices are new and being developed at the international research centres, for example systems which use Leucaena in rows with grass between (Plate 7.22) or arable crops (Plate 7.23). Some sources of further information on agroforestry are listed with the general references for this Chapter.

Plate 7.22 A sylvi-pastoral experiment in India. Grass between rows of leucaena is cut for stall feeding of cattle. The soil is shallow gravel and would not support a food crop (ICRISAT)

Plate 7.23 Double rows of leucaena for mulch or fodder. An annual arable crop will be planted between the rows in the rainy season (ICRISAT)

REFERENCES

CHAPTER 1

FAO. World Agricultural Statistics. FAO Statistical Pocketbook, FAO, Rome. 1985

Hutchinson, Sir Joseph. Preface to: Resource Development in Semi-Arid Lands. Royal Society, London. 1977

Ormerod W.E. The relationship between economic development and ecological degradation: How degradation has occurred in West Africa and how its progress might be halted. J. Arid Environments, I: 357-379. 1977

Sinclair A.R.E. and Fryxell J.M. The Sahel of Africa: ecology of a disaster. Canadian Journal of Zoology, 63(5): 987-999. 1985

CHAPTER 2

Adams M.E. A development plan for semi-arid areas in Western Sudan. Experimental Agriculture 11: 277-287. 1975

Bolton F.E. and Booster D.E. Tillage, moisture conservation and water use efficiency for dryland cereal production in winter rainfall regions. In: Proc. international Symposium on Rainfed Agriculture in Semi-Arid Regions, Riverside, California, 17-22 April 1977. Cannell G.H.(ed). University of California. 1975

Breman H. and Vithol P.W.J. The Primary Production in the Sahel (PPS) Project - A Bird's-eye View. CABO, Wageningen, Netherlands. 1984

Bryan R. and Yair A. (eds). Badland geomorphology and piping. Geo Books, Norwich, UK. 1982

Charreau C., Yerroanes D.M., Fauck R., Hall A.E., Dancette C.L. and Turk K.J. Dryland farming practices: semi-arid tropics. In: Proc. Int. Symp. Rainfed Agriculture in Semi-Arid Regions, Riverside, California, April. Cannell G.H. (ed). University of California. 1977

Dennett M.D., Elston J. and Prasad P.C. Seasonal rainfall forecasting in Fiji and the southern oscillation. Agric. Meteor 19: 11-22. 1978

Dennett M.D., Rogers J.A. and Stern R.D. Independence of rainfalls through the rainy season and the implications for the estimation of rainfall probabilities. J. Climatology 3: 375-384. Dennett M.D. and Rodgers J.A. A comparison of rainfall regimes at several sites using a rainfall 1986 model: An example from Botswana. In press. 1983

El-Swaify E.S., Pathak P., Rego T.J. and Singh S. Soil management for optimized productivity under rainfed conditions in the semi-arid tropics. Advances in Soil Science No. 1: 1-64. Stewart B.A.,(ed), Springer-Verlag, New York. 1985

FAO. Arid zone hydrology. Irrigation and Drainage Paper 3. FAO, Rome. 1981

Franke R.W. and Chasin B.H. Peasants, peanuts, profits, and pastoralists. Ecologist 11: 156-168. 1981

Hall A.E., Cannell G.H. and Lawton H.W. (eds). Agriculture in semi-arid environments. Springer-Verlag, New York. 1985

ICRISAT. Farming Systems Research at ICRISAT. IARC Workshop on Farming Systems Research, Hyderabad, India, Feb. 1986

ILCA. The First Years. International Livestock Centre for Africa. Addis Ababa, Ethiopia. 1980

Johl S.S. Drought loss manageroent: strategies for minimizing the risks of drought in India. In: Proc. Expert Meeting on Management, Conservation and Development of Agricultural Resources in the ECWA Region, May 1981, Damascus, Syrian Arab Republic. Vol. 2. Pub. by UN Econ. and Social Council, New York. 1981

Jones M.J. Programme for improved land and water management systems for Southern Africa. Proc. SDCC Agric. Res. Conf. Gaborone, Botswana, February 1984. Dept. of Agric. Research, Gaborone. 1984

Kampen J. and Burford J. Production systems, soil-related constraints and potentials in the semi-arid tropics with special reference to India. In: Soil-related constraints to food production in the Tropics. IRRI, Los Banos, Philippines. 1980

Keatinge J.D.H., Dennett M.D. and Rodgers J. The influence of precipitation regime on the management of three-course crop rotations in northern Syria. J. Agric. Sci., Cambridge 104: 281-287. 1980

Kronstad W.E. The Turkish experience in increasing food production in arid and semi-arid lands. In: Part A of Advances in food-producing systems for arid and semi-arid lands. Manassah, Brickey, Davenport, and Hagen (eds). Academic Press, New York. 1981

Lowdermilk W.C. Conquest of the land through 7000 years. Agricultural Information Bulletin 99. USDA-SCS. Washington DC. 1975

Rainey R.C. Rainfall: scarce resource in "opportunity country". In: Resource Development in Semi-arid Lands. pp 1-16. Royal Society, London. 1977

Sharon D. The spottiness of rainfall in a desert area. J. Hydrology, 17: 161-175. 1972

Sinclair J. Persistence of rain£all patterns through the season in Botswana. Bulletin of Agric. Res. in Botswana, No. 3. Gaborone. 1985

Stewart J.I. and Hash C.T. Impact of weather analysis on agricultural production and planning decisions in the seroi-arid areas of Kenya. J. Applied Meteorology 21: 477-494. 1982

Thalen. D.C.P. Discussion on "Fxperience in the Middle East" by V.C. Robertson, p. 96. In: Resources Development in Semi-arid Lands. Royal Society, London. 1977

Valentin C. Organisations pelliculaires superficielles de quelque sols de region subdesertique. Agadez, Rep. de Pdiger. Collection Etudes et Theses. ORSTOM, Paris. 1985

Vogel H. Terrace farming in the Yemen Arab Republic. Traditional forms of soil and water conservation and their present degradation; a case study of the Manakhah Region. Paper presented to IVth Int. Conf. on Soil Conservation, Maracay, Venezuela, 3-9 Nov. 1985. 1985

CHAPTER 3

Aldrick J.M., Howe D.F. and Dunlop C.R. Report on lands o£ the Ord River Catchment, Northern Territory. Animal Husbandry and Agriculture Branch, Dept. of Northern Territory, Australia, Tech. Bull. 24. 1985

Chambers R. Rapid rural appraisal: rationale and repertoire. Public Administration and Develop ment 1: 95-106. 1981

Collinson M. A low-cost approach to understanding small farmers. Agricultural Administration 8: 433-450. 1981

FAO. Agro-ecological zones project. World Soil Resources Report 48. FAO, Rome. 1978

Floret C., le Floch E., Romane F. and Pontanier R. Dynamique de systèmes écologiques de la zone aride. Acta Oecologica 2(3): 195-214. 1981

Keech M.A. The evaluation of erosion change using aerial photographs. Paper presented to Annual Meeting of Brit. Soc. Soil Sci. at Cambridge, UK. 1985

LRDC. The natural resources of Northern Nigeria. Land Resources Development Centre, Tolworth Tower, Surbiton, Surrey. 1968

LRDC. The natural resources of Lesotho. Land Resources Development Centre, Tolworth Tower, Surbiton, Surrey. 1968

Mann R.D. The complexity of problem identification among small-holder farmers in Africa. NCAE Silsoe/Project Report No. 3. Intermed. Tech. Dev. Group. London. 1974

Pickup G. and Nelson D.J. Use of Landsat radiance parameters to distinguish soil erosion, stability and deposition in arid central Australia. Remote Sensing of the Environment 16(3): 195-209. 1984

CHAPTER 4

1. General references on soil conservation in semi-arid areas

FAO. Soil conservation for developing countries. Soils Bulletin 30, FAO, Rome. 1976

FAO. Conservation in arid and semi-arid zones. Conservation Guide 3, FAO, Rome. 1976

Narayana V.V. Dhruva and Ram Babu. Soil and water conservation in semi-arid regions of India. Central Soil and Water Conservation Research and Training Institute, Dehradun, India. 1985

Stewart B.A., Unger P.W. and Jones O.R. Soil and water conservation in semi-arid regions. In: Soil Erosion and Conservation. Ei-Swaify S.A., Moldenhauer W.C. and Lo A. (eds). Soil Cons. Soc. Amer., Ankeny, Iowa. 1985

2. References in text

Billy B. Water harvesting for dryland and floodwater farming on the Navajo Indian reservation, 3-7. In: Rainfall Collection for Agriculture in Arid and Semi-Arid Regions. Dutt G.R., Hutchinson C.F. and Garduno M.A. (eds). Workshop at Tucson, Arizona, USA. Pub. by Commonwealth Agric. Bureau, Farnham Royal, Bucks. UK. 1981

Boersma L. and Jackson T.L. Fundamentals of water conservation practices used in the summer fallow system of farming. In: Proc. Int. Symp. on Rainfed Agriculture in Semi-Arid Regions, Riverside, California, April 1977. 1977

Bonsu, M. Organic residues for less erosion and more grain in Ghana. In: Soil Erosion and Conservation. El-Swaify, Moldenhauer and Lo (eds). Soil Cons. Soc. Amer., Ankeny, Iowa. 1985

Charman P.E.V. (ed). Conservation Farming. Soil Conservation Service of New South Wales, Australia. 1985

Collett J. and Boyd J. Eight simple surveying levels. Ag. Equipment leaflet 42. Int. Tech. Pubs. London. 1977

Doolette J.B. The application of the Australian lay farming system in North Africa. In: Proc. Int. Symp. on Rainfed Agriculture in Semi-Arid Regions. Riverside, California, April 1977. 1977

Fitzgerald K. Management and regeneration of degraded catchments and eroded pastoral land, with particular reference to range seeding. Conservation Guide 3, FAO, Rome. 1977

Hallam G.M., Van Campen W. and Roose E. Reacting to farmers' complaints of agriculturally intensive farms in southern Mali: from fixed answers to flexible response. Paper to IVth Int. Conf. on Soil Conservation, Maracay, Venezuela, Nov. 1985. Hallam G.M. and Roose E. Runoff, erosion, and soil degradation as constraints for rural development 1985 in Sudano-Sahelian areas of Southern Mali: Analysis of two conservative approaches. Paper to IVth Int. Conf. on Soil Conservation, Maracay, Venezuela, Nov. 1985. 1985

Heusch B. 50 Years of soil conservation in North Africa: a review. Proc. l0th Int. Conf. on Agric. Eng., Sept. 1984, Budapest, I, 1/6, 50-57. 1985

Hudson N.W. Soil Conservation. 2nd Edition, Batsford, London and Cornell Univ. Press, Ithaca, NY. 1981

ICARDA. Annual Report, I982. ICARDA, Aleppo, Syria. 1982

ILCA. Livestock research and food production in Africa. Annual Report 1984, ILCA, Addis Ababa, Ethiopia. 1985

Jones M.J. Conservation systems and crop production. Paper presented to Conf. on Agric. Development in Drought-prone Africa. Royal Commonwealth Soc. London. 13 Sept. 1985. 1985

Lindhauer M.G. Effect of potassium on water use efficiency. In: Nutrient balances and the need for fertilizers in semi-arid and arid regions. Int. Potash Inst., Berne, Switzerland. 1983

Manichon H. Experimental bases of dry farming techniques in arid and semi-arid zones. In: Nutrient balances and the need for fertilizers in the semi-arid and arid regions. 211-222. Int. Potash Inst., Berne, Switzerland. 1983

Mannering J.V. and Fenster C.R. What is conservation tillage? J. Soil and Water Conserv. 38(3): 141-143. Special ed. on Conservation Tillage. 1983

Muchiri G. and Gichuki F.N. Conservation tillage in semi-arid areas of Kenya. In: Soil and Water Conservation in Kenya. Occas. Paper 42, Inst. for Development Studies, Univ. of Kenya. 1983

Oertli J.J. Adaptation of plants to limited water availability. In: Nutrient balances and the need for fertilizers in semi-arid and arid regions. Irit. Potash Inst., Berne, Switzerland. 1983

Pereira H.C., Wood R.A., Brzostowski H.W. and Hosegood P.H. Water conservation by fallowing in semi-arid tropical East Africa. Empire J. Exptl. Agric. 26: 215-228. 1958

Rawitz E., Morin J., Hoogmoed W.B., Margolin M. and Etkin H. Tillage practices for soil and water conservation in the semi-arid zone. I. Management of fallow during the rainy season preceding cotton. Soi1 and Tillage Research 3: 211-231. 1983

Roose E. and Piot J. Runoff erosion, and soil fertility restoration on the Mossi Plateau (Central Upper Volta). In: Challenges in African Hydrology and Water Resources. Proc. Harare Symp., July 1984. IAHS Pub. 144. pp 485-498. 1984

Stocking M.J. and Peake L. Crop yield losses from the erosion of alfisols. Tropical Agriculture 63(1): 41-45. 1986

Swaminathan M.S. Rainfall and dryland agriculture. Proc. Symp. on Rainwater and Dryland Agriculture. IARC, New Delhi, Oct. 1980. 1982

Thomas D.B., Barber R.G. and Moore T.R. Terracing of cropland in low rainfall areas of Machakos District, Kenya. J. Agric. Eng. Res. 25: 57-63. 1980

Thomas R.G. Ancient spate irrigation in Wadi Beihan, People's Democratac Republic of Yemen. FAO, Rome. 1982

Virmani S.M. Climatic approach to transfer of farming systems technology in the semi-arid tropics. In: Development and Transfer of Technology for Rainfed Agriculture and the SAT Farmer. Int. Symp., Aug. 1979, ICRISAT, India. 1979

Walker P.J. Cropping and soil conservation in semi-arid western New South

Wales (Australia). J. Soil Conservation Service of New South Wales 38(2): 49-56, July 1982. 1982

Whiteman P.T.S. Moisture conservation by fallowing in Botswana. Experimental Agriculture 11: 305-314. 1975

Willcocks T.J. Tillage requirements in relation to soil type in semi-arid rainfed agriculture. J. Agric. Eng. Res. 30: 327-336. 1984

Wittmuss H. and Yazar A. Moisture storage, water use, and corn yields for seven tillage systems under water stress. Proc. ASAE Conf. on Crop Production with Conservation in the 80's. Dec. 1980, Chicago, Illinois. 1981

Wright P. La gestion des eaux de ruissellement. Oxfam, Ougadougou, Burkina Faso. 1984

Yadav R.C. Note on the effect of mulches on the conservation of soil moisture and on maize yield under semi-arid conditions. Central Arid Zone Res. Inst., Jodhpur, Rajasthan, India. Indian J. of Agric. Sci. 44(4): 241-242. 1974

CHAPTER 5

1. General references on water conservation and run-off farming (See also the general references for Chapter 6)

Boers T.M. and Ben-Asher J. Harvesting water in the desert. In: Annual Report 1979, International Institute for Land Reclamation and Improvement (ILRI) Wageningen, Netherlands. 1979

Cannell G.H. (ed). Proceedings of an International Symposium on Rainfed Agriculture in Semi-Arid Regions, Riverside, California, April 1977. 1977

Matlock W.G. and Dutt G.R. A Primer on Water Harvesting and Run-off Farming. 2nd Ed. Agric. Eng. Dept., College of Agriculture, University of Arizona. 1986

National Academy of Sciences. More Water for Arid Lands. Board of Science and Technology for International Development, Commission on International Relations. NAS, Washington DC. 1974

Pacey A, with Cullis A. Rainwater Harvesting - the Collection of Rainfall and Run-off in Rural Areas. Intermediate Technology Publications, London. 1986

UNEP. Rain and Stormwater Harvesting in Rural Areas. Ed. UNEP, Tycooly International, Dublin. 1983

2. References in text

Ahn P.M. Soil factors affecting rainfed agriculture in semi-arid regions, with particular reference to the Sahel Zone of Africa. In: Proc. Int. Symp. on Rainfed Agriculture in Semi-Arid Regions, Riverside, California, April 1977. 1977

ASAE. A series of papers on tests of conservation bench terracing in six western States of USA. Trans. Am. Soc. Agric. Eng., St. Joseph, Michigan. 1968

Bennett J. Tigrayan Test-bed. World Water, May 1985. 1985

Bilbro J.D. and Hudspeth E.B. Furrow diking for basin tillage. Experiment Station,- Lubbock. World Farming, 221, 7, 48: 104-105. Texas, USA. 1979

Billy B. Water harvesting for dryland and floodwater farming on the Navajo Indian Reservation 3-7. In: Rainfall Collection for Agriculture in Arid and Semi-arid Regions. Dutt G.R., Hutchinson C.F. and Garduno A. (eds). Workshop at Tucson, Arizona, USA. Pub. by Commonwealth Agric. Bureau, Farnham Royal, Bucks. UK. 1981

Brahmbatt B.M. and Patel A.S. Role of moisture conservation practices for semi-arid condition of Gujarat. Gujarat Agri. Univ. Res. J. 1983

Carr M.K.V. Land use and irrigated agriculture in the Woito and Lower Omo river valleys. UNDP/FAO Project ETH/78/013 Development of Irrigated Agriculture. 1979

Charnock G. Learning the hard way at Baringo (Kenya). World Water, May 1985. 1985

Cormack R.M.M. The causes and remedies for waterlogging of crops. Rhodesia Agricultural Bulletin No. 1719. 1954

Clark R.N. and Jones O.R. Furrow dams for conserving rainwater in a semi-arid climate. Proc. ASAE Conf. on Crop Production with Conservation in the 80' s. Chicago, Illinois, Dec. 1980. 1981

Cullis A. Small Steps Forward. World Water, May 1985. 1985

Cunningham G.M. Furrowing aids revegetation at Cobar despite the worst drought on record. J. Soil Water Conserv. of N.S.W. 23(3): 192-202. 1967

Cunningham G.M. Waterspreading, Cobar, Byrock District. J. Soil Water Conserv. of N.S.W. 31(2): 82-94. 1975

Dagg M. and Macartney J.C. The agronomic efficiency of the NIAE mechanized tied ridge system of cultivation. Exp. Agr. 4: 279-294. 1968

Damagnez J. Mediterranean climates, the economic use of water, and crop production. In: Soils in Mediterranean Type Climates and their Yield Potential, pp 43-55. Proc. l4th Colloq. Int. Potash Inst. Berne, Switzerland. 1979

DLFRS. Fifth annual report of the Dryland Farming Research Scheme (Phase III). Agr. Res. Stn. Sebele, Botswana. 1984

Dudal R. Dark Clay Soils of Tropical and Subtropical Regions. FAO Agr. Dev. Paper No. 83, FAO, Rome. 1965

El-Swaify S.A. Conservation-effective farming systems for the semi-arid tropics. ICRISAT, Hyderabad, India. 1983

El-Swaify S.A., Pathak P., Rego T.J. and Smith S. Soil management for optimized productivity under rainfed conditions. In: Advances in Soil Sci. Vol. I. Stewart (ed). 1985

EMBRAPA (Brazilian Enterprise for Agricultural Research). Pequena Irrigaçao para o tropico semi-arido vazantes e capsulas porosas. Bulletin de pesquisa No. 3, Jan. 1981. Agricultural Research Centre for Semi-arid Tropics (CPATSA), Petrolina, Brazil. 1981

Evenari M., Shanan L. and Tadmor N. The Negev, The Challenge of a Desert. 2nd Edition, Harvard Univ. Press, Cambridge, Mass. 1982

Fairbourn M.L. Field evaluation of micro-watershed and vertical mulch systems. In: Proc. Water Harvesting Symposium, Phoenix, Arizona, March 1974. Frasier G.W. (ed). USDA-ARS-W-22. 1975

FAO. Equipment and methods for tied ridge cultivation. Farm Power and Machinery Informal Working Bulletin 28. FAO, Rome. 1966

FAO. Arid zone hydrology. Irrigation and Drainage Paper 37. FAO, Rome. 1981

Finkel M. Report on Water Harvesting in Turkana, Kenya. Field Document·FAO-AGO-GCP/Kenya/048/NOD. 1984

Finkel M. Turkana, Water Harvesting Manual. Prepared by Finkel and Finkel, Consulting Engineers, Israel, for a training course in Kenya sponsored by FAO, NORAD, and TIDP. 1985

Flug M. Production of annual crops on microcatchments. Soils, Water and Eng. Dept., Univ. of Arizona, Tucson, Arizona, USA. CAB Farnham Royal, Bucks, UK. 1981

Gardner H. An analysis of the efficiency of micro-watershed systems. In: Proc. Water Harvesting Symposium, Phoenix, Arizona, March 1974. USDA-ARS-W-22. 1975

Garduno M.A. Research methodologies for in situ rain harvesting in rainfed agriculture. In: Rainfall Collection for Agriculture in Arid and Semi-arid Regions, pp 43-47. Dutt, Hutchinson and Garduno (eds). CAB Farnham Royal, Bucks, UK. 1981

Gindel I. Irrigation of plants with atmospheric water within the desert. Nature 207: 1173-5. 1965

Hauser V.L. and Cox M.B. Evaluation of Zingg conservation bench terraces. Ag. Eng. 43: 462-464, 467. 1962

Hurni H. Third Progress Report (Year 1983), Soil Conservation Research Project. Univ. of Berne, Switzerland. 1984

Hillman F. Water Harvesting in Turkana District, Kenya. Pastoral Network Paper l0d. ODI, London. 1980

Honisch 0. Water conservation in three grain crops in the Zambesi valley. Dept. Agric. Lusaka, Zambia. 1973

ICRISAT. Farming Systems Annual Reports 1975-1984. ICRISAT, Hyderabad, India. 1975-84

IFAD. Soil and water conservation in sub-Saharan Africa: Issues and Options. Centre for Development Cooperation Studies, Free Univ. Amsterdam, for IFAD, Rome. 1986

ILCA. Increasing African food production: ILCA's Highlands programme proposes new approach. Newsletter 4, 3 July 1985. Addis Ababa, Ethiopia. 1985

ILCA. Livestock research and food production in Africa. Annual report 1984, ILCA, Addis Ababa, Ethiopia. 1985

Jain B.L. and Singh R.P. Run-off as influenced by rainfall characteristics, slope and surface treatment of micro-catchments. Central Arid Zone Res. Inst., Jodhpur, India. 1980

Jones M.J. Conservation systems and crop production. Paper presented to Conf. on Agric. Development in Drought-prone Africa. Royal Commonwealth

Society, London, Septerober 1985. 1985

Jones O.R. Land farming effects on dryland sorghum production in the Southern Great Plains. J. Soil Sci. Soc. Am. 45: 606-611. 1981

Jones O.R. and Shipley J.L. Economics of land levelling for dryland grain production. J. Soil and Water Cons. 30: 177. 1975

Kampen J. Watershed management and technology transfer in the semi-arid tropics. Int. Symp. on Development and Transfer of Technology for Rainfed Agriculture and the SAT Farmer. ICRISAT, Hyderabad, India. 1979

Kampen J., Hari Krishna J. and Pathak P. Rainy season cropping on deep vertisols in the semi-arid tropics - effects on hydrology and soil erosion. Tropical Agricultural Hydrology. Lal R. and Russell E.W.J. (eds). Wiley. 1981

Kolarkar A.S., Murthy K.N.K. and Singh N. Water harvesting and run-off farming in arid Rajasthan. Indian J. Soil Cons. 8(2): 113-119. 1980

Kolarkar A.S., Murthy K.N.K. and Singh N. Khadin - a method of harvesting water. J. of Arid Environments 6: 59-66. 1983

Kovda V.A. Land use development in the arid regions of the Russian plain, the Caucasus, and Central Asia. In: A history of land use in arid regions. Arid Zone Research Vol XVII. UNESCO, Paris. 1961

Krantz B.A., Kampen J. and Russell M.B. Soil management differences of Alfisols and Vertisols in the semi-arid tropics. In: Diversity of Soils in the Tropics. Amer. Soc. of Agron., S. Sci. Soc. Am., ASA Sp. Pub No. 34, pp 77-95. 1978

Krantz B.A. Water conservation, management, and utilization in semi-arid lands. In: Advances in Food Producing Systems for Arid and Semi-arid Lands, Part A, pp 339-378. University of California, Davis. 1981

Lal H. Animal-drawn wheeled tool carrier: An appropriate mechanization for improved farming systems. Agricultural Mechanization in Asia, Africa, and Latin America, 16(1): 38-44. 1985

Lal H. Development of appropriate mechanization for the "W"-form soil management system. Soil and Tillage Research 8, Special Issue Nov. 1986, pp 145-160. 1986

Lawes D.A. Rainfall conservation and the yield of cotton in Northern Nigeria. Empire J. Experimental Agriculture 29(116): 307-318. 1961

Lawes D.A. A new cultivation technique in Tropical Africa. Nature, 198: 1328. 1963

Lawes D.A. Rainfall conservation and the yields of sorghum and groundnut in Northern Nigeria. Experimental Agriculture, 2: 139-146. 1966

Lyle W.M. and Dixon O.R. Basin tillage for rainfall retention. Trans. ASAE 20: 1013-1017. 1977

Macartney J.C., Northwood P.J., Dagg M. and Lawson R. The effect of different cultivation techniques on soil moisture conservation and the establishment and yield of maize at Kongwa, Central Tanzania. Trop. Agric. Trinidad 48: 9-23. 1971

Mickelson R.H., Cox M.B. and Musick J. Runoff water spreading on levelled cropland. J. Soil and Water Cons. 20(2): 57-60. 1965

Monteith J.L. Dew. Quart. J. Royal Meteorological Soc. 83: 322-341. 1957

Morgan W.T.W. Sorghum gardens in South Turkana. Geographical Journal 140: 80-93. 1974

Morin G.C.A. and Matlock W.G. Desert strip farming - computer simulation of an ancient water harvesting technique. US Agric. Res. Stn. Service, Western Region ARS-W Agric. Res. Serv. US Dep. Agric. Res. Serv. US Dep. Agric. 22, 141-150. Feb. 1975. 1975

Morin J., Rawitz E., Hoogmoed W.B. and Benyamini Y. Tillage practice for soil and water conservation in the semi-arid zone. III.Run-off modeling as a tool for conservation tillage design. Soil and Tillage Research 4: 215-224. 1984

Naghan G.P. Ecology of floodwater farming in N. America. Agro-ecosystem 5: 245-255. 1979

Narayana V.V. Dhruva and Ram Babu. Soil and water conservation in semi-arid region of India. CSWCRTI. Dehradun, India. 1985

Neff E.L. Snow trapping by contour furrows in southeastern Montana. USDA, SEA-AR, Northern Plains Soil and Water Research Center, Montana, USA. J. of Range Management 33(3): 221-223. 1980

Nilsson Gosta I. A strip-ploughing method, suggested for trial on dryland farms. Unpub. Memo, Botswana.

Njihia C.M. The effect of tied ridges, stover mulch, and farmyard manure on water conservation in a medium potential area, Katumani, Kenya. In: Soil Tillage and Crop Production, pp 295-302. Lal R. (ed). IITA, Ibadan, Nigeria. 1979

Pacey A., with Cullis A. Rainwater Harvesting - the Collection of Rainfall and Run-off in Rural Areas. Intermediate Technology Publications, London. 1986

Pathak.P., Miranda S.M. and El-Swaify S.A. Improved rainfed farming for semi-arid tropics - implications f or soil and water conservation. In: Soil Erosion and Conservation, pp 338-354. E1-Swaify, Moldenhauer and Lo (eds). S. Cons. Soc. Am. 1985

Phillips R.L. Surface drainage systems for farm lands (Eastern United States and Canada). Transactions Am. Soc. Agric. Eng. 6: 313-317, 319. 1963

Quilty J.A. Soil conservation structures for marginal arable areas: Diversion spreader banks and tank drains. J: Soil Water Cons. of N.S.W. 28,(4) 157-168. 1972

Rawitz E., Morin J., Hoogmoed W.B., Margolin M. and Etkin H. Tillage practices for soil and water conservation in the semi-arid zone. I. Management of fallow during the rainy season preceding cotton. Soil and Tillage Research 3: 211-231. 1983

Robertson L.A.N. Report on single-line tile drainage experiments in Southern Rhodesia. Res. Bull. 16. Dept. of Cons. and Extension, Harare, Zimbabwe. 1964

Robinson J.B.D., Brook T.R. and de Vink H.H. A cultivation system for groundwater soils, Part 1. E. African Agric. J. 21: 69. 1955

Ryan J.G., Sarin R. and Pereira M. Assessment of prospective soil, water, and crop management technologies for the semi-arid tropics of peninsular India. Proc. Int. Workshop on Socioeconomic Constraints to Development of Semi-Arid Tropical Agriculture, ICRISAT, Hyderabad, India. 19-23 Feb. 1979. 1979

Saulmon R.W. Snowdrift management can increase water-harvesting yields. J. Soil Water Cons. 28(3): 118-121. 1973

Shanan L. and Tadmor N.H. Micro-catchment systems for arid zone development. Centre Int. Agric. Cooperation, Min. of Agriculture, Rehovot, Israel. 1979

Sims D. Pnudivales, a practical and effective way of reducing winter waterlogging and preventing soil erosion on cereal land. Project Working Paper, Drainage and Soil Cons. in the Alentejo Region of Portugal, POR/81/002. FAO, Rome. 1983

Smith P.D. and Critchley W.R.S. The potential of run-off harvesting for crop production and range rehabilitation in semi-arid Baringo. In: Soil and Water

Conservation in Kenya. Proc. Second National Workshop, Nairobi, Kenya, March 1982. Occ. Paper 42, Inst. for Devel. Studies, Univ. of Nairobi. 1983

Stern P.H. Small-scale Irrigation. Intermediate Technology Pubs., London. 1979

Stewart B.A., Dusek D.A. and Musick J.T. A management system for the conjunctive use of rainfall and limited irrigation of graded furrows. Soil Sci. Soc. Am. J. 45: 413-419. 1981

Stewart B.A., Unger P.W. and Jones O.R. Soil and water conservation in semi-arid regions. In: Soil Erosion and Conservation. E1-Swaify S.A., Moldenhauer W.C. and Lo A. (eds). Soil Cons. Soc. America. 1985

Swindale L.D. and Miranda S.M. The distribution and management in dryland agriculture of vertisols in the semi-arid tropics. Proc. Symp. on the Properties and Utilization of Cracking Clay Soils, August 1981. In: Reviews in Rural Science No. 5, 1984, Univ. of New England, Australia. 1984

Tregubov P.S. Effective erosion control in the USSR. In: Soil Conservation: Problems and Prospects, p 454. Morgan R.P.C. (ed). Wiley. 1981

Vogel H. Terrace farming in the Yemen Arab Republic. Traditional forms of soil and water conservation and their present degradation, a case study of the Manakhah Region. Paper presented to IV Int. Conf. Soil Cons. Maracay, Venezuela, Nov. 1985. 1985

Warwick D.A. Annual Reports 1978/79 and 1979/81, Dryland Crop Agronomy Section (Big Bend). Agric. Res. Div., Min. of Ag. Swaziland. 1979/80

Willcocks T.J. The tillage of clod-farming sandy loam soils in the semi-arid climate of Botswana. Soil Tillage Research 1: 3223-350. 1981

Wollen L. Furrow cultivation (sowing) of cereals in Poland. Akademia Rolnicza, Krakow, Poland. Postepy Nauk Rolniczych, 1974, No. 4: 17-26. 1974

Wright P. La gestion des eaux de ruissellement. OXFAM, Ougadougou, Burkina Faso. 1984

Zingg A.W. and Hauser V.I. Terrace benching to save potential run-off for semi-arid land. Agron. J. 51: 289-292. 1959

CHAPTER 6

1. General references on water harvesting and use

Burton J. Water Storage on the Farm. Water Research Foundation of Australia. 1965

Cooley K.R., Dedrick A.R. and Frasier G.W. Water harvesting: state of the art. In: Watershed Management. Proc. Symp. 11-13 August 1975. 1975

Dutt G.R., Hutchinson C.F. and Garduno M.A. Rainfall Collection for Agriculture in Arid and Semi-Arid Regions. Proc. Workshop sponsored by University of Arizona, USA and University of Chapingo, Mexico, at Tucson, Arizona, Sept 1980. Published by Commonwealth Agricultural Bureaux, Farnham Royal, UK. 1981

FAO. Water laws in Moslem countries. Irrigation and Drainage Paper 20/1. FAO, Rome. 1976

FAO. Watershed Management Field Manual. FAO Conservation Guide, Volume 13/2 Water Harvesting. FAO, Rome. 1986

Frasier G.W. ed. Proceedings of Water Harvesting Symposium. Phoenix, Arizona, March 1974. USDA-ARS-W-22. 1975

Frasier G.W. and Myer L.E. Handbook of Water Harvesting. Agricultural Handbook 600, USDA-AR6. 1983

Hollick M. Water harvesting in the arid lands. Scientific Reviews on Arid Zone Research Vol. 1: 173-247, Scientific Publishers, Jodhpur, India. 1982

Hudson N.W: Field Engineering for Agricultural Development. Oxford Univ. Press. 1975

Kennedy W.K. and Rogers T.A. Human and animal-powered water lifting devices. Intermediate Technology Pub., London. 1985

Nissen-Petersen E. Rain Catchment and Water Supply in Rural Africa: A Manual. Hodder and Stoughton, London. 1982

2. References in text

Chiarella J.W. and Beck W.H. Water harvestinng cacahments on Indian lands in the Southwest. In: Proc. Water Harvesting Symposium, Phoenix, Arizona, March 1974. Frasier G.W. (ed). USDA-ARS-W-22, pp 104-111. 1975

Cluff C.B. Engineering aspects of water harvesting research at the Univ. of Arizona. In: Proc. Water Harvesting Symposium, Phoenix, Arizona, March 1974.

Frasier G.W. (ed). USDA-ARS-W-22, PP 27-39. 1975

Cluff C.B. Surface storage for water-harvesting agrisystems. Proc. Workshop sponsored by University of Arizona, USA and University of Chapingo, Mexico, at Tucson, Arizona, Sept 1980. Published by Commonwealth Agricultural Bureaux, Farnham Royal, UK, pp 23-30. 1981

Cooley K.R. Evaporation suppression for conserving water supplies. In: Proc. Water Harvesting Symposium, Phoenix, Arizona" March 1974. Frasier G.W. (ed). USDA-ARS-W-22. 1975

Crow F.R. and Manges H.L. Comparison of chemical and non-chemical techniques for suppressing evaporation from small reservoirs. Trans. ASAE 10(2): 172-174. 1967

Dedrick A.R., Hansen T.D. and Williamson W.R. Floating sheets of foam rubber for reducing stock tank evaporation. J. Range Man. 26(6): 404-406. 1973

Dept. of Agriculture, Western Australia. A method for optimising cost of dam and roaded catchment combination. Technote 4/76 DAMCAT. 1979

Dept. of Agriculture, Western Australia. Roaded catchment design and construction. Farmnote 109. Roaded catchments: maintenance. Farmnote 129. 1984

Dutt G.R. Establishment of NaCl-treated catchments. Proc. Workshop sponsored by University of Arizona; USA and University of Chapingo, Mexico at Tucson, Arizona, Sept 1980. Published by Commonwealth Agricultural Bureaux; Farnham Royal, UK. 1981

Evenari M., Shanan L. and Tadmor N. The Negev, The Challenge of a Desert. 2nd Edition, Harvard University Press, Cambridge, Mass. 1982

Fink D.H. Paraffin-wax water-harvesting soil treatments improved with antistripping agents. Soil Science 138(1): 46-53. 1984

Frasier G.W. and Cooley K.R. Performance evaluation of water harvesting catchments. J. Range Management, 32, 6, Nov. 1979. 1979

Frith J.L. Design and construction of roaded catchments. In: Proc. Harvesting Symposium, Phoenix, Arizona, March 1974. Frasier G.W. (ed). USDA-ARS-W-22, pp 122-127. 1975

Frith J.L. Design standards for farm surface water supplies. Wheatbelt Water Supply Seminar. Water Res. Foundation of Australia, Northam WA. 1977

Frith J.L., Nulsen R.A. and Nicol H.I. A computer model for optimizing design of improved catchment. In: Proc. Water Harvesting Symposium, Phoenix, Arizona, March 1974. Frasier G.W. (ed). USDA-ARS-W-22, pp 151-157. 1975

Hanson C.L., Neff E.L. and Woolhiser D.A. Hydrologic aspects of water harvesting in the northern Great Plains. In: Proc. Water Harvesting Symposium, Phoenix, Arizona, March 1974. Frasier G.W. (ed). USDA-ARS-W-22, pp 129-140. 1975

Hollick M. The design of roaded catchments for maximum run-off. In: Proc. Water Harvesting Symposium, Phoenix, Arizona, March 1974. Frasier G.W. (ed). USDA-ARS-W-22, pp 210-220. 1975

Hollick M. Water harvesting in the arid lands. Scientific Reviews on Arid Zone Research 1: 173-247. Scientific Publishers, Jodhpur, India. 1982

Houston J. Rapid relief for Zimbabwe. World Water, May 1985, p 31. 1985

ITDG. The introduction of rainwater catchment tanks and micro-irrigation to Botswana. Intermediate Technology Pubs., London. 1969

Karan F. and Julka R. The use of soda-clay in reducing seepage in water harvest technology (Storage and conservation methods, India). Trans. Ind. Soc. Desert Technology and Univ. Centre of Des ert Studies. 6,1, Jan. 1981. 1981

Laing I.A.F. Sealing leaking excavated tanks on farms in Western Australia. In: Proc. Water Harvesting Symposium, Phoenix, Arizona, March 1974. Frasier G.W. (ed). USDA-ARS-W-22, pp 159-174. 1975

Laing I.A.F. Rainfall collection in Australia. Proc. Workshop sponsored by University of Arizona, USA and University of Chapingo, Mexico, at Tucson, Arizona, Sept 1980. Published by Commonwealth Agricultural Bureaux, Farnham Royal, UK., pp 61-66. 1981a

Laing I.A.F. Evaluation of small catchment surface treatments. Aust. Water Res. Council, Tech. Paper 61. Canberra. 1981b

Laing I.A.F. and Pepper R.G. Sealing farm dams. J. Agric. Western Australia 17(2): 2-7. 1976

Laing I.A.F. and Prout A.L. Bitumen oil, and clay surfaces on a deep sand to increase run-off from catchments for excavated tanks in Western Australia. In: Proc. Water Harvesting Symposium, Phoenix, Arizona, March 1974. Frasier G.W. (ed). USDA-ARS-W-22. 1975

Laing I.A.F., Pepper R.G. and Prout A.L. Flat batter dams suit difficult sites. J. Agric. Western Australia 4: 110-112. 1980

McBride M.W. and Shiflet L.W. Water harvesting catchments in the Safford District, south eastern Arizona. In: Proc. Water Harvesting Symposium, Phoenix, Arizona, March 1974. Frasier G.W. (ed). USDA-ARS-W-22, pp 115-121. 1975

Mickelson R.H. Performance and durability of sheet metal, butyl rubber, asphalt roofing, and bentonite for harvesting precipitation. In: Proc. Water Harvesting Symposium, Phoenix, Arizona, March 1974. Frasier G.W. (ed). USDA-ARS-W-22. 1975

Myers L.E. and Frasier G.W. Asphalt-fibreglass for precipitation catchments. J. Range Management 27(1): 12-15. Jan. 1974. 1974

Odell M.L. and Odell M.J. The evolution of a strategy for livestock development in the communal areas of Botswana. Pastoral Network Paper l0b, ODI, London. 1980

Plueddemann E.P. Soil treatment with silicones for water harvesting. In: Proc. Water Harvesting Symposium, Phoenix, Arizona, March 1974. Frasier G.W. (ed). USDA-ARS-W-22, pp 76-83. 1975

Poesen J. Surface sealing as influenced by slope angle and position of simulated stones in the top layer of loose sediments. Earth Surface Processes and Landforms 11: 1-10. 1986

Pressland A.J. and Lehane K.J. Run-off and the ameliorating effect of plant cover in the Mulga communities of south western Queensland. Aust. Rangeland J. 4(1): 16-20. 1982

Richards K.S. Rain-water harvesting for domestic purposes. In: Water in Agriculture. Rhodesian Agric. J. Tech. Bull. 15, pp 45-51. 1972

Roose E. and Piot J. Run-off erosion, and soil fertility restoration on the Mossi Plateau (Central Upper Volta). In: Challenges in African Hydrology and Water Resources. Proc. Harare Symp., July 1984. IAHS Pub. 144, pp 485-498. 1984

Sandford S. Management of Pastoral Development in the Third World. Wiley, Chichester, UK. 1983

Shanan L, Morin Y. and Cohen M. A buried membrane collector for harvesting rainfall in sandy areas. Proc. Workshop sponsored by University of Arizona, USA, and University of Chapingo, Mexico, at Tucson, Arizona, Sept 1980. Published by

Commonwealth Agricultural Bureaux, Farnham Royal, UK. pp 67-76. 1981

Sheikh M.I., Shah B.H. and Aleem A. Artificial catchments for rainwater harvesting in deserts of Pakistan. Pakistan For. Inst., Peshawar, Pakistan. 32(1): 7-17. 1982

Sinclair A.R.E. and Fryxell J.M. The Sahel of Africa: ecology of a disaster. Canadian J. Zoology 63(5): 987-999. 1985

Tadmor N.H. and Shanan L. Run-off inducement in an arid region by removal of vegetation. Proc. Soil Sci. Soc. Amer. 33(5): 790-793. 1969

Valentin C. Organisations pelliculaires superficielles de quelques sols de region subdesertique. AGADEZ, Rep. de Niger. Collection Etudes et Theses, ORSTOM, Paris. 1985

Wade N. Sahelian drought: no victory for western aid. Science (Washington DC) 185: 234-237. 1974

Welchert W.T. and Freeman B.N. Horizontal Walls. Jo Range Management 26(4): 253-256. 1973

Wulff H.W. The qanats of Iran. Scientific American 218: 94-101. 1968

CHAPTER 7

1. Some general references for 7.1 - Grazing land

Alchin B.M. Species trials in semi-arid south western New South Wales, 1956-1971. J. Soil Cons. Serv. New South Wales 30(1): 46-73. 1974

Palit S.B. Species for wood, food, and fodder in arid and semi-arid zones of India. In: Sand dune stabilization, shelterbelts, and afforestation in dry zones. Conservation Guide 10, pp 183-188. FAO, Rome. 1985

2. Some general references for 7.2 Trees and shrubs

Anon. A strategy on the role of forestry in combatting desertification. Paper presented at the Int. Arid Lands Research and Development Conference - Arid Lands Today and Tomorrow - Tucson, Arizona, Oct. 1985. 1985

Armitage F.B. Irrigated Forestry in Arid and Semi-Arid Lands: A Synthesis. IDRC Publication 234E, Ottawa. 1985

Berry L. Agro-forestry in the West African Sahel. Advisory Committee on the Sahel. Nat. Res. Council, Washington, USA. 1984

Chandler T. and Spurgeon D. International Cooperation in Agro-forestry.

ICRAF, Nairobi, Kenya. 1979

Lundgren B. and Nair P.K.R. Agro-forestry for soil conservation. In: Soil Erosion and Conservation. E1-Swaify, Moldenhauer, and Lo (eds). pp 703-717. Soil Cons. Soc. Amer., Ankeny, Iowa. 1985

Nair P.K.R. Agro-forestry Species: a crop sheets manual. Pub. 003E. ICRAF, Nairobi, Kenya. 1980

3. References in text

Anderson R. Imprinting technique offers new scope for reforestation. Ceres 19(3): 3-5. 1986

Barrow E.G.G. Use of microcatchment for tree planting in soil conservation in semi-arid areas. In: Soil and Water Conservation in Kenya. Thomas D.B. and Senga W.M. (eds). Inst. Dev. Studies and Fac. of Agric. Univ. Nairobi, pp 32.4-332. 1983

Boers T.M. Economic evaluation of water harvesting in microcatchments. Water Resources Research 19(5): 1099-1105. 1983

Brown L.R. Reversing Africa's Decline. Worldwatch Paper 65, Washington DC. 1985

Cunningham G.M. Regeneration of scalded duplex soils in the Coolabah district of New South Wales. J. Soil Cons. Serv. of New South Wales 30(3): 157-169. 1974

Dixon R.M. Land imprinting for controlling infiltration and desertification processes. ASAE Paper 83-2514. 1983

Dragoun F.J. and Kuhlman A.R. Effect of pasture management practice on run-off. J. Soil Water Cons. 23(2): 55-57. 1968

Eckholm E.P. Losing Ground: Environmental Stress and World Food Prospects. Norton, New York. 1976

Erickson L.R. and Currie P.O. A multifunction rangeland improvement machine for semi-arid regions. ASAE Paper 82-1021. 1982

Evenari M., Shanan L. and Tadmor N. The Negev, The Challenge of a Desert. 2nd Edition. Harvard University Press, Cambridge, Mass. 1982

Fitzgerald K. Management and regeneration of degraded catchments and eroded pastoral land with particular reference to range reseeding. In: Conservation in arid and semi-arid zones. Conservation Guide 3, pp 41-60, FAO, Rome. 1976

Fortmann L. and Roe E. Dam groups in Botswana. Pastoral Network Paper 12b, ODI, London. 1981

Fowler M. Overgrazing in Swaziland. A review of the technical efficiency of the Swaziland herd. Pastoral Network Paper 12d, ODI, London. 1981

Gallacher R. Niger - Conservation des eaux et des sols. FAO, Rome. 1974

Hadley R.F. Evaluation of land-use and land-treatment practices in semi-arid western United States. US Geol. Survey, Denver, Colorado, USA. Philosophical Transactions of the Royal Society of London, B. 1977, 278, 962, 543-554. 1977

Harrington G.N., Wilson A.D. and Young M.D. Management of Australia's Rangelands. Div. of Wildlife and Rangelands Research, CSIRO. 1984

Hays T.R. Modern and ancient utilization of rainfall in the Sudanese Butana (abstract) Assoc. for Arid Lands Studies, Proc. Vol. IV. Publ: Lubbock, Texas, USA. 1981

Houerou H.N. le. Agroforestry techniques for the conservation and improvement-of soil fertility in arid and semi-arid zones. ILCA, Addis Ababa, Ethiopia. 1980

Jones R.M. Scald reclamation studies in the Hay District of New South Wales: Part III. Reclamation by ponding banks. J. Soil Cons. Serv. of N.S.W. 23(1): 57-71. 1967

Keetch R.I. Rangeland rehabilitation in Central Australia. Land Cons. Unit, Conservation Commission of the Northern Territory. 1981

Kowsar A. Water harvesting for afforestation. II. Dependence of tree growth on amount and distribution of precipitation. J. Soil Sci. Soc. Amer. 46(4): 802-807. 1982

Martin C.S. and Ward D.E. Rotating access to water to improve semidesert cattle range near water. J. Range Management 23(1): 22-26. 1970

Neff E.L. and Wright J.R. Contour furrowing as a range management practice. In: Proc. Symp. on Watershed Management, Boise, Idaho, July 1980. 2, 827-835. ACSE, New York. 1980

Newcombe K. An Economic Justification for Rural Afforestation: the Case of Ethiopia. Energy Department Paper 16, World Bank, Washington DC. 1984

Odell M.L. and Odell M.J. The evolution of a strategy for livestock development in the communal areas of Botswana. Pastoral Network Paper l0b.

ODI, London. 1980

Ormerod W.E. The relationship between economic development and ecological degradation: How degradation has occurred in West Africa and how its progress might be halted. J. Arid Environments 1: 357-379. 1978

Rawitz E. an.d Hillel D. Water harvesting by run-off inducement for irrigation of an almond orchard in a semi-arid climate. In: Proc. Water Harvesting Symposium, Phoenix, Arizona, March 1974. Frasier G.W. (ed). USDA-ARS-W-22, pp 223-232. 1975

Shanan L. and Tadmor N.H. Micro-catchment systems for arid zone development. Centre of Int. Agric. Cooperation, Min. of Agriculture, Rehovot, Israel. 1979

Simanton J.R., Osborn H.B. and Reynard K.G. Effects of brush to grass conversion on the hydrology and erosion of a semi-arid southwestern rangeland watershed. Hydrology and Water Resources in Arizona and the Southwest 7: 249-256. 1977

Stocking M.A. Development projects for the small farmer: Lessons from eastern and central Africa in adapting conservation. In: Soil Erosion and Conservation. E1-Swaify, Moldenhauer and Lo (eds). Soil Cons. Soc. Amer., Ankeny, Iowa. 1985

Tromble J.M. Semi-arid rangeland treatment and surface run-off. J. Range Management 29(3): 251-255. 1976

UNDP. Project completion report, Eastern Senegal Livestock Development Project. Western Africa Projects Dept. (WAPAC), World Bank, Washington DC. 1984

Valentine J.F. Range Development and Improvements. Brigham Young University Press, Utah, USA. 1971

Viertmann W. Amenagements agro-sylvo-pastoraux dans les bassins versants du centre et du nord-ouest Tunisien. Final Report on Watershed Management TUN/81/004. FAO, Rome. 59p. 1984

ANNEX 1:

GLOSSARY OF LOCAL NAMES

AHAR: Northern India, earth contour bund to retain floodwater, Section 5.3.3.

ATAJADIZO: Mexico, silt-trap dam for water storage in the sediment, Section 6.3.3.

CORDONS: planted Francophone West Africa, stone contour lines, maybe with vegetation, Section 7.2.2.

CUVETTE: Francophone West Africa, shallow basins which collect surface run-off for cereal or fodder production in the basin, Section 5.3.1 and 7.1.2.

DIGUE: Mali, large flood diversion bund with control gates, Section 5.3.4.

DIGUETTE: Burkina Faso, semi-circular small walls of earth or stone to retain surface run-off for improved cropping of cereals or fodder, Section 7.1.2.

FANYA JUU: Kenya, literally 'throw uphill', a kind of terracing where soil is excavated and thrown uphill to form a bank" Section 4.3 and Figure 4.2.

GESSOUR (also Jessour): Tunisia, terraced valley bottoms, Section 5.3.2.

GILGAI: Australia, broken uneven surface of clay soils with humps and hollows, Section 5.4.

HAFIRS: Arabic-speaking Africa, excavated water storage tanks, Section 5.4.

KAIR: USSR, farming of seasonally flooded river bank terraces, Section 5.3.1.

KHADIN: Northern India, Rajasthan, earth dam to retain floodwater for nundation and cropping of reservoir floor, Section 5.3.3.

KHAKI: Turkmenistan, USSR, run-off farming with floodwater, Section 5.3.1.

KHARIF: India, the summer monsoon season, Section 5.3.3.

KHUL: Northern India and Pakistan, irrigation canals cut into steep valley sides, Section 5.3.2.

KUSKABA: Pakistan, inundation farming using floodwater, also SAILABAS, Section 5.3.3.

LIMAN: USSR, run-off farming by spreading floodwater with large earth banks, Section 5.3.2.

LUNETTE: Francophone Africa, shallow half -moon depressions for tree planting, Section 7.2.2.

MESKAT: Tunisia, run-off farming of olives, using shaped catchment areas and conducting channels, Section 5.3.2.

MULGA: Australia, local name for Acacia aneura, an unpalatable shrub, Section 5.3.2.

MURUNDUM: Brazil, large earth banks for interception and storage of storm run-off on arable lands, Section 4.3.1 and 5.3.2.

NEGARIN: Israeli term, derived from 'neger' meaning run-off, and applied to several different types of surface shaping into small basins or microcatchments, Section 7.2.2.

PANI PANCHAYAT: India, a local water management committee for all water affairs Section 1.1.

PNUDIVALE: Portugal, shallow surface drains; with slight ridge formation, on arable lands with mechanized farming, Section 5.4.

PYNE: India, flood diversion canals above river bed level to divert high stage floods for inundation farming, Section 5.3.4.

QANAT: Iran and other countries in northern Africa and western Asia. Also called 'karez', 'foggara', and 'falaj'. Horizontal wells, Section 6.5.2.

RABI: India, the dry post-monsoon winter cropping period, Section 5.3.3.

SAILABA: Pakistan, inundation farming using floodwater, also called kuskabas, Section 5.3.3.

SAWAGI: Yemen Arab Republic, supplementary irrigation using collected run-off on rainfed cultivated terraces, Section 5.3.2.

SAYL: Yemen Arab Republic, irrigation by floods diverted from wadis onto levelled, cultivated terraces, Section 5.3.2.

TERA: Sudan, basins behind earth bunds for inundation cropping, Section 5.3.3.

TRINCHERA: Latin America, silt-trap dam for water storage in the sediment, Section 6.3.3.

VAZANTE: Brazil, areas inundated by run-off stored behind semi-circular earth banks, Section 5.3.3.

WADI: Arabic-speaking Africa, usually dry streambed, with intermittent flash floods, Section 5.3.2.

WARABUNDI: India and Pakistan, a locally managed irrigation water committee - pukka warabundi with a formal structure based on legislation, katcha warabundi without formal structure, Section 1.1.

WARPING: China, irrigating with sediment-laden water to maintain fertility, Section 5.3.4.

ZAI: Burkina Faso, small excavated basins or 'water pockets', similar to 'cuvettes' Section 5.3.1.

Table 2.7/7. Summary of FAO–Unesco system of soil classification, with analogues in the U.S. system of taxonomy

FAO–Unesco system			Analogues in present U.S. system of taxonomy
Symbol	Soil unit	Description	
J	Fluvisols	Soils of alluvial deposits and floodplains	Fluvents
G	Gleysols	Soils genetically dominated by excess of water	
	eutric, dystric		Haplaquents, Psammaquents, Tropaquents, Andaquepts, Fragiaquepts, Haplaquepts, Tropaquepts
	mollic		Haplaquolls
	humic		Humaquepts
	plinthic		Plinthaquepts
	gelic		Pergelic cryaquepts
R	Regosols	Soils on loose material with weak or no development	Orthents, Psamments
I	Lithosols	Soils with hard rock at shallow depth	Inceptisols
Q	Arenosols	Weakly developed, open, coarse-textured soils	Psamments
E	Rendzinas	Shallow humus-carbonate soils	Rendolls
U	Rankers	Shallow humus-silicate soils	Haplumbrepts
T	Andosols	Soils formed from materials rich in volcanic glass and commonly having a dark, humic surface horizon	Andepts
V	Vertisols		Vertisols
	pellic		Pelluderts, Pellusterts, Pelloxerts
	chromic		Chromuderts, Chromusterts, Chromoxererts, Torrerts
Z	Solonchaks	Saline soils	Aquepts, Orthids
	orthic		Salorthids
	mollic		Salorthidic Calciustolls, Salorthidic, Haplustolls
	gleyic		Halaquepts
S	Solonetz	Sodic soils	Alfisols, Aridisols, Mollisols
	orthic		Natrustalfs, Natrixeralfs, Natrargids, Nadurargids
	mollic		Natralbolls, Natriborolls, Natrustolls, Natrixerolls
	gleyic		Natraqualfs
Y	Yermosols	Desert Sand Plain soils	Aridisols
	haplic		Camborthids, Durorthids
	calcic		Calciorthids
	gypsic		Gypsiorthids
	luvic		Argids
X	Xerosols	Semi-desert soils	Molic Aridisols
	haplic		Mollic (xerollic or ustollic) Camborthids and Durorthids
	calcic		Mollix (xerollic or ustollic) Calciorthids
	gypsic		Mollix (xerollic or ustollic) Calciorthids
	luvic		Mollic (xerollic or ustollic) Haplargjds and Durargids
K	Kastanozems	Soils rich in organic matter having a brown or chestnut colour	Ustolls
	haplic		Haplustolls, Aridic Haploborolls
	calcic		Calciustolls, Aridic Calciborol
	luvic		Argiustolls, Aridic Argiborolls
C	Chernozems	Soils rich in organic matter having a black colour	Ustolls
	haplic		Haploborolls, Vermiborolls
	calcic		Calciborolls
	luvic		Argiborolls
H	Phaeozems	Soils rich in organic matter having a dark colour	Ustolls
	haplic		Hapludolls
	calcic		Vermudolls
	luvic		Argiudolls
	gleyic		Argiaquolls

Table 2.7/7. (Cont'd)

FAO–Unesco system			Analogues in present U.S. system of taxonomy
Symbol	Soil unit	Description	
M	Greyzems	Soils rich in organic matter having a grey colour	Ustolls
	orthic		Argiborolls
	gleyic		Aquolls
B	Cambisols	Soils weathered in situ having a cambic horizon	Ochrepts, Tropepts
	eutic		Eutrochrepts, Ustochrepts, Xerochrepts, Eutropepts
	dystic		Dystrochrepts, Dystropepts
	humic		Haplumbrepts, Humitropepts
	gleyic		Aquic Dystrochrepts or Eutrochrepts
	calcic		Eutrochrepts, Ustochrepts, Xerochrepts
	chromic		Xerochrepts
	vertic		Vertic Tropepts
	ferralic		Oxic Tropepts
L	Luvisols	Soils with an illuvial accumulation of clay in the B horizon	Alfisols
	orthic		Hapludalfs, Haploxeralfs
	chromic		Rhodoxeralfs, Haploxeralfs
	vertic		Vertic Haploxeralfs
	ferric		–
	albic		Eutroboralfs
	plinthic		Plinthustalfs, Plinthoxeralfs
	gleyic		Aqualfs
B	Podzoluvisols	From Podzols and Luvisols	Boralfs, Aqualfs
	eutric		Glossoboralfs
	dystric		–
	gleyic		Glossaqualfs, Aquic Glossaqualfs, Aquic Glossoboralfs
P	Podzols		Spodosols
	orthic		Orthods
	leptic		–
	ferric		Ferrods
	humic		Humods
	placic		Placorthods, Placohumods
	gleyic		Aquods
W	Planosols	Soils developed in flat terrain mostly with poor drainage	
	eutric		Abaqualfs, Paleargids, Palexeralfs, Paleustalfs
	dystric		Albaquults
	mollic		Mollic Albaqualfs, Argialbolls
	humic		–
	solodic		–
A	Acrisols	Acid soils with low base saturation	Ultisols
	orthic		Hapludults, Haplustults, Haploxerults
	ferric		Palexerults, Paleustults
	humic		Humults
	plinthic		Plinthaquults, Plinthudults, Plinthustults
	gleyic		Aquults
N	Nitosols	Soils showing a shiny ped surface	Alfisols, Ultisols
	eutric		Propudalfs, Paleudalfs, Rhodustalfs
	dystric		Tropudults, Rhodudults, Rhodustults, Palexerults
	humic		Tropohumults, Palehumults
F	Ferralsols		Oxisols
	orthic		Orthox, Torrox, Ustox
	xanthic		Orthox, Torrox, Ustox
	humic		Humox
	acric		Acrox
	plinthic		Plinthaquox